VIKING TO VICTORIAN

BY

OLAF T. ENGVIG

OLAF T. ENGVIG

VIKING TO VICTORIAN

EXPLORING THE USE OF IRON IN SHIP BUILDING

THEMO PUBLISHING
LOS ANGELES
2006

Prepress & graphics Nancy Wansick, Hercules, CA
Paper: 157 gsm Art paper gloss
Set in Book Antiqua
Printed in China by Kong Tai Printing Ltd.
First Printing 2006
Published in the USA

* * *

The author has received financial support from the
Norwegian Non-fiction Literature Fund

Craft Press Inc, San Francisco
THEMO Publishing
Los Angeles

ISBN 978-0-9655451-6-7
ISBN 0-9655451-6-4

Library of Congress Cataloging-in-Publication Data

Includes bibliographical references and index.
History, technology, industrial development, iron production, shipbuilding,
sailing, travel, transportation, laboratory research and analysis.
Book design by Mona and Olaf T. Engvig
Published in the United States of America

A vast number of coastal and ocean going vessels have been constructed through times with the use of wood and iron.
Few of them still exist in a relatively original condition.
Those remaining vessels should be kept intact.
All entities in charge of such vessels should protect them by overseeing that the original hull will not be harmed or replaced.
Such acts will lead to the permanent loss of unique historic information that should be saved for future generations to interpret.

HM The King of Thailand's Royal Barges are kept in a building and out of the water.
They are only used on special occasions. Artist unknown.

*"It is a good generation that saves for its children things to see and learn from,
to make them feel important and proud of themselves and their ancestors' achievements."*

- "Shipping and Culture," San Francisco 1997, p. 131

This book is dedicated to those who strive to save,
restore and work to achieve the same rights
for old vessels as for other historical landmarks.

First and foremost:

Lars Bache Nielsen
Ole Haugerudbråten
Mona Beichmann
Knut Petterson
Knut Lein
Atle Valland
Stig Sølvberg
Erling Beenfeldt
Johan Aase
Kristian Korsmo
Arne Lannerstedt
and
Morten Hjelløy

CONTENT

November 20, 2000

Foreningen HANSTEENS VENNER
c/o Knut Petterson
Oslo Byrett
Pb. 8023 Dep.,
0030 Oslo
NORWAY

Dear Sirs:

As Executive Director and Chief Curator of the San Diego Maritime Museum, caretaker of the 1863 barque *Star of India* (formerly *Euterpe*), I have been asked to give a statement about the uniqueness of the iron in S/S *Hansteen*.

First, let me say that both of these vessels represent a significant waypoint in the development of shipbuilding and indeed, in the history of technology. Beginning in the middle years of the 19th century, the advent of iron construction applied to a wide variety of manufacture produced enormous change in the modern world and signaled the rise of large scale technological systems spanning the globe. In some cases iron allowed the first creation of technological systems, such as railroads, which not only generated enormous leaps in transportation, but established individual regions and nations on trajectories of technological development with far reaching consequences. In others, such as shipbuilding, iron gradually penetrated and then supplanted far more ancient methods of manufacture with similar results. Worldwide, the advent of iron manufacture produced one of the great technological revolutions of history.

Yet, for all its significance, wrought iron manufacture in large scale technological systems flourished for only a brief period before it was supplanted by steel produced through the Bessemer process. By the last quarter of the 19th century, most large or complex structures formerly built of wrought iron were now manufactured of steel because of its comparative strength and economy of production. In consequence, there are relatively few large scale artifacts from the wrought iron industrial period that have survived. Because ships historically suffer enormous service life attrition, they are the rarest of all. One such vessel, *Star of India*, continues to sail today and has been formerly recognized by the World Ship Trust (1996) as a treasure of humanity. Significantly, all components of her original hull remain intact after 137 years of salt water emersion, 21 circumnavigations, and many encounters with storms, reefs, ice, fire, collision, and neglect. Though not so light as steel for the same strength, her iron has proven to be an extraordinarily durable material and there is no reason to doubt that this ship will continue to go to sea for decades or even centuries more without necessity for significant replacement of her original historic fabric.

Hansteen is a similar example of iron shipbuilding and an artifact of enormous historical importance. I have read Olaf T. Engvig's report on the iron of S/S *Hansteen*, and am very saddened to learn that her iron hull has been disturbed since her restoration was completed. I certainly hope that the remaining part of her iron construction can be kept intact, so the ship can be saved for future generations as a unique document of the golden age of wrought iron ship construction. Since there are so very few wrought iron ships left in the world in seaworthy condition, Norway has a unique opportunity to make a mark internationally in relation to maritime heritage.

It is my strong belief that the few remaining iron ships should not be altered, but kept and maintained in a manner that does not further damage or hurt their original historic fabric. Only by preserving the iron in S/S *Hansteen*, will future generations have a unique historic vessel to experience and to research.

Yours Sincerely,

Raymond Ashley
Executive Director,
San Diego Maritime Museum

INTRODUCTION

When I started restoring the research ship *Hansteen* from 1866, I noticed that the old wrought iron plates of her hull seemed to be in better condition than some plates that obviously had been replaced. Over the years, while I worked on this ship and other old ships, I became more and more aware of the great difference between iron and steel. I had samples from the plating of the *Hansteen* sent to a laboratory for metallurgical testing and analysis (Nyrud, 1982). The results were so encouraging that I collected and sent more samples (Johnsen, 1984). The conclusion strongly indicated what I suspected: that old hammer iron, or wrought iron, seemed much more resilient as a ship building material than the more modern steel products that had been used for replacements. My earlier experience with iron rivets in the lap-streaks of old traditional Norwegian longboats was similar. In this flexible construction, the iron rivets were usually the last part to give in.

I then started on a small crusade to save old iron. I knew that in construction of buildings, bridges, ships and railroads, the use of wrought iron declined sharply in the 1880s. After 1890, most shipyards were building their tonnage of steel. The same happened in the building industry and the railroads. The latter abandoned iron rails after Sir Thomas Brassey, the great railway entrepreneur, wrote the obituary on wrought iron rail in 1879.

Today, old iron constructions still left in the world are rare historical landmarks that people have paid little attention to, apart from structures we see as masterpieces. Ship preservationists seem more determined to do riveting repairs with modern steel in old iron structures to make them look old and genuine, rather than trying to save the iron in the original riveted structures. Riveting was used until the 1950's and later in many types of steel construction, surviving wrought iron construction by more than 70 years.

There are major differences between a riveted wrought iron structure and a structure of steel, although they may look similar. Construction methods and riveting techniques were not the same, but most significantly, the iron production process was fundamentally different, and so was the finished product. Steel repairs and alterations to old wrought iron structures like a ship's hull can be devastating. Once the wrought iron is discarded, the relic loses an important part of its historic integrity, as a riveted iron construction is a complete item and an authentic part of the past. With professional care, adequate protection and careful use, the last iron ships might outlast any ship made of organic material, or steel for that matter, leaving us with documentation of old construction techniques, skilled workers' and craftsmen's work, tool marks, and other specific signatures for new generations to enjoy, research and interpret.

Relatively few original wrought iron structures are left untouched. They should be preserved. So should ships, even if sometimes groundings and other damages during a vessel's long and active duty make repairs with steel necessary. These repairs are valid because they tell us a story.

Over-restoration is an ongoing issue, often approved and supported financially by authorities. It is very late, but we may still be able to protect some genuine iron ships from an important period of invention and development.

A major part of our planet's surface is covered with water. Floating constructions of any kind have always been important to men. During the Viking Age, wrought iron rivets made from bog iron ore kept ocean-going Viking ships together. Norsemen used a combination of lap-streak ship construction and iron rivets to construct light and swift sailing ships that could even be rowed. In those open boats of wood, they traveled far, because their ships had the ability to take a lot of punishment and stay afloat on the high seas. At the same time, those vessels were capable of negotiating inshore waters, as well as rivers and lakes, due to their lightness and extreme shallow draft. Expeditions to inland environments demanded a swift vessel that under any condition of wind and weather, could be good enough to overcome swift river currents, carried across land if necessary, as well as function as a protective vehicle against hostile attack from native tribes. If the Vikings could not overpower such attacks, they could usually take to the oars, or hoist the sail and be gone before the enemy could get them.

The Viking Age shipbuilders used iron and wood to develop an optimal low-weight and high-strength ship that could be a superb platform for transport on any occasion and even crossing the most dangerous of all seas—the wet and cold North Atlantic Ocean. Only when those ships became old, or encountered extreme weather or ice, would they be lost at sea. The Hansa-cogs and similar decked vessels developed in Europe during the Medieval Age were able to make crossings in the wake of the Vikings, but were hardly suitable for navigating inland rivers and lakes with shallow water with steady currents that force the vessels off-course.

The use of iron in the construction of wooden ships would take place when the strength of the metal was required—only then would its use be justified. In the hundreds of years after the Viking Age, this would be the rule. Only when industrialized production of iron started on a larger scale, would the use of more iron in ship building pave the way for new adaptations. The world witnessed the development of small steam ships built of iron using a modification of the old lap-streak construction. They soon started to grow in size. For a brief period, the construction of composite (wood on iron) vessels became popular, soon to be relieved by new ships with entire hulls built of iron with decks of wood placed on iron beams.

Although thousands of wrought iron vessels were constructed during the latter part of the 19th Century, few have been left in their original states. The salt in the sea made the iron more susceptible to rust than structures on land. Also, collision, wind and weather, changing currents, fog and shifting sands would end the working days of many good ships before they got very old. If a ship was among the few lucky ones that managed to survive well into old age, it always had residual value to the owner. Usually it would be moved to a breaker that would pay handsomely to break it up for the metal to be used in new products. A ship had a value as long as it existed and could be collected and recycled. Of all types of wrought iron structures, the ship was among the most common constructions. Today, it is among the rarest.

Fortunately, some ships have managed to survive these challenges to become genuine survivors from this first era of modern ship construction. But regrettably, some owners tend to restore and rebuild the unique iron ships with modern steel plates and angels, arguing that it is the same material, that the old iron looks bad, and that it does not have the same strength as new plates. Some restorers even say that "it is just rusty old iron and nothing to save." (O. Haugerudbraaten, personal communication, 2000)[1].

Today, restored iron vessels from the mid-19th Century are presented as originals, when they are in fact, steel ships. Many have been rebuilt and restored excessively. What those ships tell us today is not necessarily a good story on how a historic ship is protected. The most regrettable part is that those vessels are lost to genuine research and information. We do have examples of so-called genuine ships that are less than 10% original and should be re-classified. It would be better to build a true replica and to save the old iron hull. By doing that, we leave something original, real and unique for generations to come. Heavily restored iron ships should have a label with some kind of benchmark for originality.

No genuine Viking ship exists today without considerable restoration and replacement of deteriorated or missing material. The first generation of iron built sailing vessels and steamers have witnessed much of the same development. Instead of leaving the structures as intact as possible, they were repaired and renewed to give the best impression of how they were when new. If the objective is to re-create the vessel, would a complete new replica not serve the world much better?

I believe it is time to accept that Disneyland and other theme parks, as well as replica ships, are good and educational places for an experience of something that has been lost. However, the most genuine artifacts should be left in a protective environment such as a museum or an institution for future generations to deal with and investigate.

---

[1] Haugerudbraaten communicated with the leader of the shipyard, certified to work on historic ships, on the ethics of replacing old iron with steel, when he was informed that some 20 plates had been abandoned and would be replaced with steel in the hull of the *Hansteen*. He was told flat out: "Why keep it? It is just rusty old iron."

My research and testing since the 1970s on the iron rivets in open Norwegian longboats dating back more than a hundred years, the stamp marks in the iron ship *Hansteen* and the *Star of India*, and several other iron vessels from the 19th Century has given me a whole new understanding of the significance of wrought iron in ship construction. Inquiries on stamp marks found on different ship building profiles and plates made in the 1860s and the production of such at named ironworks in Great Britain and Norway have stirred up a lot of new interest for old hammered wrought iron.

Only few years ago, most scholars considered old iron artifacts—often with pits and partly damaged by rust—the least interesting object to be saved and fully agreed that it should be replaced using steel. (Even welding was argued to be the only option in restoring old ships according to museum authorities some years back.) This is all changing. With the salvaging of the *Hunley*, the world's first submarine to sink a ship, and the recovery of the turret and other iron artifacts from the *Monitor*, the Americans have engaged the world in saving and protecting old iron for the future, as the British have, in restoring the *Great Britain* and the saving of the iron ship the *Warrior*.

It is very important to stop vandalizing old iron boats, bearing in mind that the "golden age" of iron ship building only lasted for about one generation. Even so, it was the generation that launched new developments and expanded the world with steamships through the Suez Canal and railways across America. That iron period paved the way for even better and bolder achievements. The story of the production and use of industrial iron is important for the understanding how all this happened.

The excellent condition of the iron plates in the hull of the *Hansteen* triggered my curiosity and made me continue my research into iron in ship building. I decided to keep this ship as original as possible to show the world how well the oldest generation of ships tonnage has survived. After 15 years, and as the first completely restored major historic ship in Norway, and by far the most original one of that age, the *Hansteen* was finished and a commemorative building plaque was made.

FROM VIKINGSHIP TO IRON SHIP

NORSEMEN AND THE NORTHERN SEAS

A thousand years ago, people from Norway ventured deep into the vast northern sea, sailing to the British Isles, Ireland and the islands north and west of Scotland, and further on to the Faeroes, Iceland, Greenland and America. These voyages were so common that only some odd ventures that went astray were mentioned in the Sagas. So were enterprises to new lands as part of a re-settling movement. Regular commercial transport to and from Iceland, Greenland, Vinland and other destinations in the North Atlantic, and in the Arctic Sea to Svalbard for fishing and hunting, would be considered uneventful and hardly worth mentioning. Voyages to America after the Greenlanders' attempt to settle there were not recorded even when we know they built at least one community in Newfoundland for many years and continued south along the American coast and rivers. Little was known of these activities, but records tell us that they must have had good and seaworthy vessels that could take on long and enduring voyages across the Atlantic Ocean on a regular basis.

While the Portuguese and Spaniards ruled the Southern Seas mostly by coastal explorations, many nameless Norsemen conquered the Northern Seas by crossing vast areas of open, cold and hostile waters where others would not dream of staking out a course, and they did it with simple ways of navigation in open boats of wood and iron completely exposed to the elements of nature.

Few of the Norse navigators have been named, but among those few are two world famous expedition leaders that represent the beginning and the end of the age of great explorers. The European expansion of their world by exploring the rim and beyond the parameter of known land took off at about AD 1000 when Leif Eiriksson landed on America and stayed for the winter. Nine hundred years later, Roald Amundsen was the first man to sail through the Northwest Passage — a task people had tried to accomplish for 300 years. He also went to the South Pole, the North Pole and the North East Passage. Amundsen was the last of the great explorers that discovered and mapped unknown parts of the world at a time when man knew much more about the surface of the moon than the polar regions of our own planet. Amundsen collected the first knowledge of the vast white areas in what might be named: "The conquest of the last wilderness of the earth's surface." All the great explorers on sea and land came between Eiriksson and Amundsen.

Throughout history, nations and individuals have competed to find new land to establish colonies or hunt for gold and silver. Individuals obtained fame and admiration. Geography and anthropology became popular sciences. Eiriksson, Marco Polo, Henry the Navigator, Vasco da Gama, Columbus, Magellan, Sir Francis Drake, Cabrillo, John Cabot, Hudson and Franklin were names known to every educated person. Governments would support expeditions that could boost the nation's image. Amundsen, like Nansen before him, obtained support for polar expeditions that were as challenging as the space program of the latter part of the 20th Century.

A common vessel for all world explorers was the ship. It was the platform from which knowledge of new lands and faraway continents was obtained. The construction of better and more lasting vessels to reach these far away destinations was a key factor in the advancement of modern day culture. Only with the use of iron could the Americans conquer the Wild West and the British build and maintain an empire that stretched all across the world.

This is not a replica but a 100% percent original working longboat built in 1863 which was never restored or renewed. This first photo of the 29 feet long *Hitra* was taken in 1976, during the fifth time it was out of the boat-shed since the 1890s.

ENCOUNTER WITH IRON

When restoring the *Hansteen,* I found that the old iron seemed to be in better shape than the replacements on the hull. It was a surprising discovery to me even when I knew that iron rivets in old open boats were resilient. To people that had worked with old wrought iron, it was hardly a surprise. When I was in York on my first Viking sail mission in 1978, and later during my many visits there, I met and had interesting talks with Chris Topp — a young, able blacksmith and a devoted student of old wrought iron who worked with York Archaeological Trust. We discussed about the old iron rivets in my original open longboat *Hitra* which was built in 1863. In spite of being used on the Norwegian coast on and off for generations, none of the many home-made iron rivets or roes had given in. The long and enduring voyages to York and across the North Sea when the *Hitra* was well over a hundred years old proved that old wrought iron is an excellent product for keeping such a vessel together. I told Chris of my experience with the hull of the survey ship *Hansteen,* built three years later than my longboat, with industrial iron. He told me that artifacts of wrought iron many hundred years old could be almost as good as new, even when exposed to wind and weather, if the iron was of good quality, (C. Topp, personal communication, 1978).

This was consistent with the knowledge I had acquired over the years. I was born and raised by the Trondheimsfjord and learned to row an open boat as soon as I could hold an oar. There were many old *faering* longboats of the Viking type still around. The last generation of these traditional boats built for their original purpose and daily use were made during a local "work project," to keep young men away from service for the German occupants during World War II. Retired boat builders came out of retirement to be instructors. *Faerings* were being built again, but only for the duration of the war. They were more seaworthy and could carry a sail and two pairs of oars that gave them a much wider range than the successor, the short-range rowboat. Small motorboats that had become common as fishing vessels during the inter-war period could not get fuel for their engines, so people switched back to their old boats for fishing and transport. Since the early 1900s, lap-streak boats had been built with machine-made ships nails, rivets and round roes. Earlier boats were riveted together with "home made" iron rivets, usually made in the boat builder's own smithy. The blacksmith boat builder would make the rivets from a piece of iron. The farms would usually also have the wood and work the forest, pick the threes, split the logs and undertake all the steps needed to make a complete boat. A good calf or a barrel of grain could be a fair price for a *faering* that could easily be made in a week with the raw materials on hand.

When I grew older, experienced fishermen told me that the old riveted boats had to be re-riveted, because the rivets had rusted away. Indeed, I was shown boats that had been re-riveted with new rivets halfway between the old rivets. I later learned that this statement could only be partly true. I came to believe that this field needed a closer investigation in determining the exact age and the builder of each boat. There were obviously a lot of misunderstandings and misconceptions. The common saying was that the new boats were superior to older ones. Often this was said without any knowledge of the boats' real age. Regardless of whether they were 20, 40 or 80 years old, they were all "old boats." I tried to classify different periods, types and quality of iron, but it was difficult and I did not have the success I had hoped for. There were a lot of errors and conflicting statements. Also, I was told that when the boat building farmer was running low on iron, he might dig out a piece from the back of his smithy, perhaps left by his grandfather, and use that, or other persons on the farm might use an old piece of scrap iron from equipment that could be recycled. It was even possible that farmers in this part of Norway might find old Viking iron or iron in the remains of a farm deserted during the bubonic plague that swept Europe in the mid-14th Century. The man that ordered the boat might bring iron for rivets if the boat builder asked for it (Engvig, 1969–75).

After interviewing shipbuilders, fishermen and other coastal residents, visiting and inspecting many old traditional crafts between 15 and 30 feet in length, I thought I had acquired a good knowledge of these Viking-type boats. On the coast of mid-Norway, these boats were in common use longer than in most other areas along the Norwegian coast and also had a closer appearance to an original boat from the Viking Age. My general conclusion was that the really old boats seemed to be in better shape than many of the younger ones, making me believe that the older iron rivets were of a better or more lasting quality. When I found the *Hitra*, a Viking-type longboat from 1863 in perfect and original condition, I knew what the old man, Reidar Melkvik, said was true. He told me that I did not have to re-rivet the more than 100-year-old boat; the iron rivets were of good quality even if they were rusty. Reidar Melkvik was right — *Hitra* was a good boat made of the best quality wood and iron (R. Melkvik, personal communication, 1976).

When I later started my work on the wrought iron sail steamship *Hansteen*, I experienced much of the same attitude from ship surveying people and shipyard workers as those I had met working with wooden boats: "Old iron is no good." Rusty, ugly and beaten up plates of iron needs to be replaced with steel to make a ship safe and certified to sail with passengers. "It makes no difference as long as you rivet" was a statement I often heard after I had reintroduced riveting as a way of restoring historic ships. Maritime offices and museums accepted my way of re-riveting instead of welding, but with some hesitation. Even Wilhelm Trygg, the old riveter I engaged, who reintroduced the art of riveting to the people in Norway on film and on TV, first tried to convince me that I should be welding. "It is much better than riveting," he said in my first interview with him (W. Trygg, personal communication, 1979).

After I came to San Francisco, I met and befriended Chief Curator Karl Kortum, the founder of San Francisco Maritime Museum. He was a well-known ships' savior — he had restored and helped saved old ships very much the same way I had done in Norway. We had much in common and many things to talk about in regard to collecting and keeping old historic ships. One important topic was iron versus steel in historic ships. He had more or less the same opinion as me: the hulls of the oldest ships were often in better general condition than younger vessels. He had been to the Falkland Islands and inspected the *Great Britain* built in the 1840s and was stunned by how well her hull had survived. He was pleased to hear of my experience and iron research on the *Hansteen* yielded the same conclusion. Iron from ironworks in Great Britain, Norway and Sweden produced around 1860 seemed to keep up much better than the 20th Century vessels made from newer and different quality iron products. "Old hammer iron and salt sea seem to work well together," I told Karl, and he agreed (K. Kortum, personal communication, 1992).

Karl Kortum (far-right) and Olaf Engvig (center) seated at their usual corner of the Buena Vista Bar and Restaurant at Hyde Street Pier, San Francisco with their favorite waitress. This was the place where they discussed their experience with wrought iron versus steel, and made new plans for the protection of old ships.

Having said that, one should adjust for the use, exposure and maintenance of the vessel, which is always important. I have seen the bottom frames of the hold of an old cemented vessel from 1891 that worked as a freighter for salted cod and herring for many years. The result speaks for itself. The upper and exposed part of the iron is completely rusted through, while the bottom of the same frame, protected by the cement, is as good as new. The lifespan of a 15,000-ton tramp motor tanker sailing clean products in the 1950s was estimated to be 11 years. Another rule of thumb for ocean going tonnage might be that older ships tended to be constructed with heavier material for the tonnage than newer vessels, leaving greater leeway for deterioration.

The *Star of India* and the *Hansteen* were both heavily constructed mainly to make them safer vessels, last longer and to get better insurance rates. The *Star of India* is a large, bark-rigged ship for ocean trade. The *Hansteen* is a smaller sail steamship built to do depth soundings for the government of Norway outside the rocky Norwegian coast, littered with many hidden rocks not marked in any sea map. It was predicted that she would ran aground, and she was constructed accordingly. Later, during her service, it was recorded with pride that the *Hansteen* ran up on submerged rocks at full speed a couple of times and got stuck, but without any serious damage to the ship's hull. During one incident, she was left almost dry during ebb tide.

While wooden ships in those days were sometimes salted inside for protection, iron ship was often cemented. The cement at the bottom served several purposes. It protected the iron very well. It also served as permanent ballast. Pig iron bars and broken cast iron pieces could be added into the cement for added ballast weight. Finally, it made the bottom stronger and could prevent the soft iron plates from being dented if the bottom should hit an object with a moderate force. For outside hull protection, the mill scale on the surface of the iron was a good start, but barnacles and other marine growth would soon settle. Early maintenance and protection of the outside hull was done in a traditional way, often by giving the iron plates a couple of coats of good pitch tar.

When the cement was removed from the plates inside the *Hansteen*, it exposed the grey iron surface that never saw rust. The plates were as clean as the day they were delivered from the ironworks in 1866. Rust was only present in the iron that has been exposed above the cement.

THE IRON AGE IN SHIP BUILDING

A thousand years before ships were made from iron, blacksmiths used wrought iron to make a wide range of items such as tools, weapons, utensils, hinges and various reinforcements. They decorated doors and made rivets and roes for the construction of a wide range of Viking ships. Many myths developed around this trade. The blacksmith was the wizard who could manipulate pieces of red-hot iron, and create wonderful, long-lasting items, which were highly prized and needed. According to Norwegian folktales, the blacksmith was the only man that could level with the devil.

The Norse people before the Vikings never had much of a Bronze Age, even if they were masters in making beautiful items of bronze. They developed astonishing skills based on their melting and casting techniques. Compared to many other areas they also had particularly good fuel for making the right heat. The Vikings were among the first people to depend on significant use of iron in their ship building. Without iron, the shipbuilder would not be able to make a Viking ship that would hold together under tough conditions.

In Northern Europe, the earliest and most common form of iron for the smithy came from bog iron ore through bloomery production. The Norsemen or Vikings that settled in America could have used the site at L'Ance aux Meadows as a base camp, or for trading and local exploring, but such assumptions have not been extensively proven.

We know that they produced good quality iron from bog iron ore at this location (A. Espelund, personal communication, March 12, 2004) and repaired or made new ships and boats. Quantities of discarded fragments of old boat rivets tell us that the iron was recycled (B. Wallace, 2001)[2]. The numerous fragments and remains of rivets and roes suggest to me that they may have come from abandoned vessels rather than repairs. Shavings also could indicate that they worked on boats. I would postulate that the Norse had a shipyard in America a thousand years ago, and that this can be supported by archaeological finds at the site. We will return to this topic later. Bloomery production of iron continued in many different places on the Scandinavian Peninsula long before Norway and Sweden became nation states.

The use of iron in boat and ship construction has been alive in that part of the world for much more than a thousand years. Around the AD 1800, it developed into the construction of complete iron ships based on mined iron ore, made into iron, first with charcoal and later by coal from coal mines. When iron ship building started on the British Isles, the conversion started from alterations to the old lapstreak or clinker built construction techniques for keeping a wooden

---

[2]Unpublished manuscript and records of numerous fragments of boat nails and rivets found during the Ingstad excavation on the site at L'Anse aux Meadows, Newfoundland, provided by Birgitta Wallace.

hull together, the common construction form since before the Viking era. Norway, in particular, has an unbroken tradition in the construction of clinker-built wooden vessels. Riveting new iron ships together was accomplished in much the same way and basically with the same tools and techniques as using rivets in a lapstreak construction. Hull plates on smaller iron vessels would often not be much broader than the boards in the hull of a boat of wood. Those making the finished iron vessel appear with the same lines along the hull as any wooden boat would have. This would gradually change as iron producing techniques and constructional methods were refined and iron ships were turned out in greater numbers and rapidly growing sizes.

England, with its vast quantities of fuel in their coal mines, became the great mass producer of hammered wrought iron. Many of the ironworks in Scandinavia that delivered iron for the first iron ships to be built locally based their iron production on coal made from wood. They could not compete, and almost all of them were closed down after 1850. The hammer iron production on the British Isles really took off and British ironworks delivered ship quality iron for a large-scale iron ship building industry at numerous shipyards along its coasts, and for export. The building of new iron tonnage in many parts of the world was made possible by the advent of British iron. A series of new inventions and more efficient steam engines made this new Iron Age possible. It far exceeded anyone's dream only few generations before. We will return to this subject later.

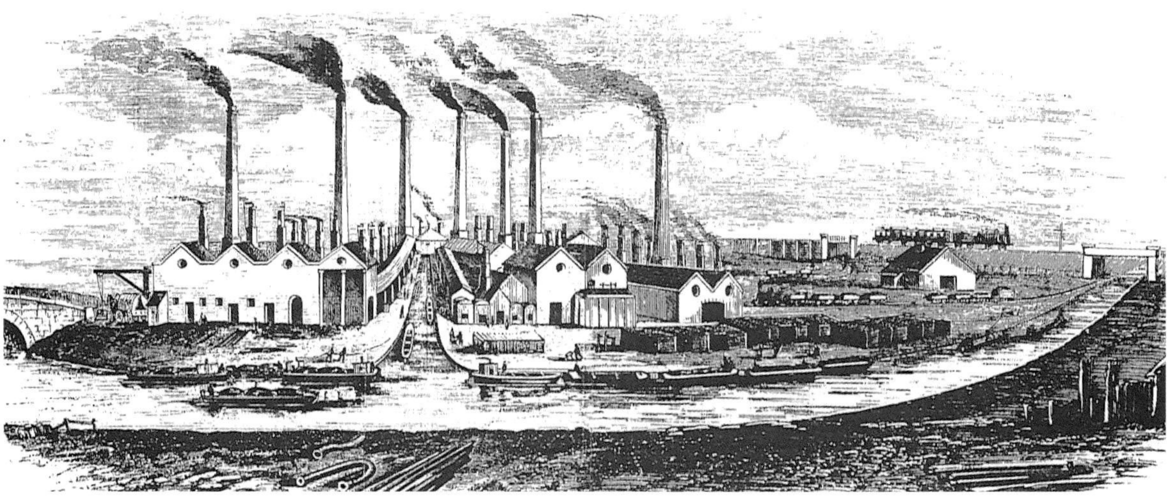

Bloomfield Iron Works in Staffordshire was one of the well-known providers of good quality iron for shipyards around the world that had engaged in iron ship building during the 19th Century. This engraving, courtesy of Tom Round, also shows a curved canal, railroad and three bridges.

Alas, this golden age of hammer iron production soon ran up against newer and bolder achievements in the field. Further discoveries and development made even better iron products available on the market at a competitive price. When ship owners discovered the steel ship's superiority in strength, they switched to steel. After 1880, many of the former prosperous ironworks were either shut down or converted into steel mills. The grand era of iron ship construction lasted only for approximately one generation. Iron ship building ended with the Victorian era. Ever since then, the world tonnage has been constructed of steel. This is why the last remaining iron ships are such historic treasures. They are the genuine introduction to our own time with regard to shipping.

What makes the old wrought iron so resilient towards deteriorating forces in the nature? According to "The Wrought Iron Advisory Centre" in England, mass-produced puddle iron owes much of its rust-proof properties to its fibrous nature. The refining of the product is a crude process, and results in the inclusion of non-corrodible slag in the structure. The outer surface of the iron is called the mill scale. The ironworks tried to make iron that did not develop rust easily on the surface. To test the iron's ability to withstand rust, newly produced iron plates and angles were taken to the ironworks back lot where it was stored in open sheds. If it could remain there for a year or more without developing rust, it could be sold as best grade iron. Official bodies oversaw the production quality. Best Yorkshire was a brand name of superior grade that only few ironworks were able to obtain (Larsstuvold, 1982). Plates and angle iron were stamped with the producing mills' signature and given the quality impressions, so as to enable construction companies to determine what material they were using for their records.

If the inside of a ships hull is well kept and maintained, it will carry the ironwork's stamp marks on its inside angel iron and deck beams. Bulkheads and the outside hull will likewise carry impressions done on every plate during production. For obvious reasons, plate marks or impressions are not common. Marks might be shallow, and wear and tear could have eroded them completely (Engvig, 1986).

There is one technique to determine if the ship's plates are an original part of a ship's hull. If the frames carry stamps of originality and the area seems untouched by repairs, one can usually assume that the plates must be the original. In most cases, repairs to a riveted iron hull can be spotted easily. In the case of the *Hansteen*, we had no problem determining which ones of the 126 plates in the ship had been repaired or renewed. This was consistent with reports and laboratory tests, which are presented later in this book.

Grit blasting was developed for the cleaning of modern steel plates. It is sometimes used for cleaning and removing paint on old hammer iron ships. This is a devastating way of trying to clean old iron. It removes the outer surface of the iron, which should be left intact, because it protects the iron from rust. On an iron ship, grit blasting of the hull will immediately blow away the iron and often results in the finding of laminar spots that the sandblaster will try to clean away. That is the code working on any ordinary ship. The result could well be a continuation of blasting until the entire 8-millimeter or more of original iron is completely bore through, ruining the iron plate that was in good condition before the grit blasting work started. Later reports would, of course, state that the plate was bad.

Wrought iron is a laminar material. There are examples of governmental surveyors, only familiar with modern steel, demanding parts of a hull plate be changed because of false readings on the ultra sound meter due to fragments of impurity inside the ship's iron plate. By using a small bit, boring these spots in the old fashioned way, readings will show if the plate is of the required thickness or not. It is too late when one discovers that the removed cut-out has the same thickness throughout, and was in perfect condition. The harm has been done and it is not possible to reverse it. This was done to the hull of the *Hansteen* after the ship had been fully restored (O. Haugerudbraaten, personal communication, 1997)[3]. A 19th Century wrought iron plate is very different from a modern certified electro steel plate, just like a natural meadow is different from a farmer's monoculture field. You cannot apply the same "tools" in both instances.

Meadows are protected in national parks, and modern machinery is kept away. The National Parks are saved for us and future generations to enjoy, and the National Park Service is set-up to protect and oversee the upkeep of these national treasures. Yellowstone and Yosemite became National Parks more than a hundred years ago. The need to establish good protection procedures for old wrought iron constructions is long overdue.

The last iron ships, as well as other old ships of wood, should be left alone. Small groups of volunteers and museums have tried to convince authorities that they should do the same with ships as they do with other relics. The case has been argued through letters, media and other means for many years. Saving and protecting old vessels has raised little interest among today's politicians and bureaucrats.

Many genuine iron ships have suffered excessive damage through restoration attempts to make them seaworthy again. The list of such examples is long. Real originality becomes more and more rare. Lack of support from national or regional authorities results in shortage of funding or loss of knowledge (or both), and is the main factor in the demise of old historic ships. It leaves the field open for all types of random development. The world has seen many saviors who gave up on their work due to lackluster support and left famous historic vessels to deteriorate. Others ships have obtained enough money and support to be saved and "restored" to a degree where the ships were licensed for sailing with tourists. It makes some money for its own upkeep, gets media attention, and gives the crew their "day on the bay."

---

[3]Report on two cut-outs of iron plates on the *Hansteen*, removed by mistake after wrong meter readings. When taken out they both proved to be good and of adequate thickness. This was reported by the leader of the yard, who regretted that had to replace them.

TWO IRON VESSELS AND A LAP-STREAK LONGBOAT

The research ship *Hansteen* on its maiden voyage and new trial run in 1993, after the ship's iron hull had being fully restored from the keel and up, and her old interior wood had been put back to where it was originally. Only seven 1866 iron plates along with all the 19 steel plates that the ship had acquired during her long-time service due to groundings and accidents were replaced during the restoration period. This was, in 1993, a unique ship with a major part of the lower hull made from iron.

A few years after *Hansteen* was completely restored and every part of the hull had been approved by Norway's Maritime Directorate's surveyor, the ship was grit blasted in spite of the fact that the new Board was strongly advised against it. That treatment destroyed more than 17 original iron plates that had to be replaced with steel plates to make her a seaworthy vessel once more. The cost of these repairs far exceeded the entire price of her full restoration from a hulk. It is doubtful if the *Hansteen* still can be labeled as a true iron ship, as more than 50% of the material in the hull now consists of steel.

The hull design of the *Hansteen* has never been changed. This ship has seen three major refittings, but only been extended upwards. But it has had several bumps with the rocky coast over the years. As a research vessel, she touched the sea bottom several times. The ship's keel shoe, bow and bottom plates still had clear proof of those incidents when we started. *Hansteen* ran so hard aground on unknown rocks that she was left dry during ebb tide. A big boat was needed to pull her off. She did not suffer a leak during any of these episodes.

Later, she had other brushes with land, and even tried to sink a big German transporter during World War II by ramming amidships. A few days later, an enemy warship tried to make payback and dented her entire starboard side above the original iron hull by broad-siding her. The added-on weather deck sustained considerable damage. The court ruled that the Germans were to blame on both occasions. The ship suffered little damage to the original hull, apart from an ugly dent in the bow that we finally straightened out.

When she was fully restored in 1993, however, her hull still mostly consisted of iron. In the early 1980s, seven original and bumpy bottom plates were replaced and the bow was repaired. The rest of the 28 plates renewed were secondary steel plates, and all of them were in a worse state of decay than any of the original wrought iron plates. The rest of her hull as of 1993 was that of the ship's original construction in 1866 as far as we know.

In order to use the *Hansteen* as a pleasure vessel, the ship's new owners, a provincial government with no knowledge in preservation of maritime iron, started cutting out "bad" parts they had discovered on her iron hull. The same hull had been cleared with the appropriate authorities in Oslo as sound material a few years before. Pitted old iron plates on the side of the ship looked less attractive than the smooth new steel plates where the ship had been repaired over the more than hundred years of service. They therefore grit blasted the hull to clean it. The result was as expected. The sandblaster found impurity in the iron, continued blasting and destroyed some twenty good old iron plates.

The state antiquarian and the shipyard assigned to do the work on the hull could see no harm in replacing old "ugly looking" iron, arguing that as long as they were riveting, it was acceptable (L.B. Nielsen, personal communication, 2002)[4]. This is a problem. Iron ship building discontinued around 1890. Riveting was suspended well after World War II, about 70 years later. What is more important: a genuine wrought iron hull or the visual impression of new rivets?

Local authorities in charge of this ship report to the state antiquarian, but they lack the means to secure an iron ship. People in charge of the summer use and maintenance on the *Hansteen* today are modern sailors with little education and knowledge about upkeeping and protecting a unique museum relic more than 130 years old.

---

[4]Also personal communication with K. Petterson, O. Haugerudbraaten and K. Lein regarding saving the *Hansteen* as an iron ship after it was fully restored. Several initiatives met with "the great silence" from media and government officials.

The *Star of India* at the Maritime Museum of San Diego. Built in 1963, she is still a true iron ship.

The *Star of India* has a very different story to tell after her sailing days as a world trader was over in the 1920s. She became a museum ship in San Diego, California only a few years after she was laid up. That happened very early compared to the bulk of museum ships of today, of which many have been saved and restored during recent years. Furthermore, the *Star of India* has been kept at this museum, and even if the ship has had its down periods, she has done very well. The natural environment in San Diego, together with her sturdy construction and able museum authorities, have helped this ship survive deteriorating forces to an amazing extent. She sets sail for a short voyage on her own about once a year for a museum appearance. She is manned by museum workers and volunteers, with modern tugboats on stand-by. The *Star of India* has a much better lease on life than the *Hansteen*, which never became part of a museum as originally planned.

The *Star of India* and the *Hansteen* are extremely interesting and valuable icons left from the classical era of wrought iron ship building. When the *Hansteen* was built in 1866, most of the world's tonnage was still being built of wood, and carried sails only. The main period of iron ship construction lasted about 35 years. This new era triggered by the industrial revolution opened up the use of iron in many new ways. It demanded new constructional techniques and was the gateway to knowledge of all developments in modern transport and communications that followed. The few fairly original iron ships left are technical marvels that should be protected and nourished in the best way possible. The iron ship period will never return, and these ships are the last source for information, interpretation and research.

In 1978, when the *Hansteen* was 112 years old, it was saved from being scuttled. It was restored to its original glory by a group of volunteers. Her hull was completely redone and approved for certification. A considerable part of her interior items were saved and reused. Her appearance when the restoring was finished in 1993 was outstanding and she was by far the most professional and scientific restoration project on a ship in Norway until then. The iron report on the *Hansteen* follows later in this book.

Further sailing of historic ships like the *Star of India* and the *Hansteen* should probably be limited in order to protect the original hull. Their usage should be adapted to the ship's state of originality. Passenger traffic should not be a guideline for preservation. Regulations and demands to meet new safety standards and other requirements are on the rise. In trying to meet these goals, the authenticity of historic ships will be compromised (as it had happened already), end up as quasi-original ships of lesser historical value.

I have seen and investigated hundreds of old lap-streak boats on the coast of Norway. When I first laid my eyes on the *Hitra*, I knew it was unique. That boat was among the oldest boats I have seen and it was in perfect condition. It had been used and kept in an excellent way. It had mainly been utilized during the cod-fishing season in the winter for many years. Marks from the hand lines were deep in the gunwales and the original sail had distinct stains of fish blood at the part closest to the hold, indicating that many a large catch had been landed. The owner had much information to share and it all corroborated with what I knew from numerous other interviews. He specified that even if it was built in 1863, it did not have to be re-riveted or repaired.

Apart from *Hitra's* age and its good condition, this particular boat was also one of the largest seaworthy boats of this longboat type I have seen. No really large Aafjord-type longboat from the time they were in commercial use has survived. I understood that this vessel was as close as I could get to the perfect descendent of the real Viking ship. Since it was built before the 1880s, the boat builder would not have had the slightest clue as to what a Viking ship really looked like. He, of course, built the boat the way he had learned. And he was known all over the region as the very best boat builder. He was the local Stradivarius and could demand extra payment for his boats. For me, this boat represented the missing link back to the Vikings and their boats.

This *stor-firing* is almost 30 feet long and only six feet across, and carried ten oars and a square sail. I was told by the owner it was in this boat that his grandfather had overtaken the 12-knot steamer that carried the "Blue ribbon" of the Trondheimsfjord more than a hundred years ago. The original sail, made in the 1860s, and all the ten oars and equipment were all preserved. Not one of the many loose parts that fit inside was missing, but a few of the oars needed replacement.

The last owner had used the boat during five winter cod-fishing seasons back in the 1920s. Since then it had been in the shed all the time until I found it in 1976. The owner wanted me to have it. When I mentioned that I would like to sail it several hundred nautical miles from the island of Hitra to Oslo, he answered that I would have no problem. From this moment, the idea of sailing "The Viking Way" started to develop in my head. I would like to show the world how this old missing link of a Viking ship would keep up on the high seas. I had been a sea cadet on Norway's large windjammer the *Statsraad Lehmkuhl* at age 15, and later a deep-water sailor on motor tankers and steam trampers, and had studied maritime history and Viking sailing at the University of Oslo for my graduate degree. Now was the time to put all my experience to the test by going to sea in a genuine and original open longboat.

The longboat *Hitra*, built in 1963, is today still an original Norwegian longboat. The interior of this 29 feet, 10 oared, vessel shows it is constructed the same way as a Viking boat from the 9th Century. The boat builder making this vessel had no idea of how a Viking ship looked like or how it was made, that information emerged later in history. It makes *Hitra* the missing link, and proves that the boat building tradition of the Vikings was kept alive at Aafjord for a thousand years. Any alteration to the ancient design would compromise its capabilities and create an inferior vessel. Again and again, the *Hitra* has proved that it is the speediest vessel under sail or oars.

The speed was great, the waves even greater. Hitra was shooting out of a big wave, which was building up behind the boat. The wind was strong. The run was more or less down wind and the speed was tremendous even with a small sail. We were lucky that the huge waves of water crests did not fall on us as we were trying to find some small island out west, without the use of a map, compass or a watch.

GOING INTO THE OCEAN

Two open boats with square sails cut through the waves, with *Hitra* leading the way. The speed was great, the waves even greater. Huge mountains of green sea water were towering up on the starboard aft quarter. The boats were rolling and yawing, even when the small sail was only half-way up the mast to help steady the vessel's movement. The people on board hung on. It was misty, and the horizon where the sky met the sea was only a couple of hundred yards ahead. The strong wind came from the northwest. The rain and spray soaked everything onboard. This sea was demanding — cold, wet and exhausting. It did not give much leeway and the people aboard has to know exactly what to do. *Hoevedsmann* and his crew were on the alert the entire time, playing hide and seek with the big waves, like the Vikings experienced in the North Atlantic Ocean a thousand years ago.

This time, the boats carried no map or compass, no watch or any other 20th Century equipment to tell where we were and where we were heading. For the first time in hundreds of years, small, open longboats propelled by a single square sail and oars alone have set sail from Norway in an attempt to reach small islands way out into the ocean in the west. It was a modest target, and the navigator was confident that he would find the way without aid.

He had been trained in Viking navigation by experienced navigators since he was a young boy. If the wind and the weather hold up, we would reach the islands in less than a week. If not, we would sail straight out into the North Atlantic, without chance of seeing any land for the several thousand miles ahead. It was a daring challenge, but we knew, like the Vikings did, that if we get lost we could always turn back. It would be land all the way where we set out. With prevailing winds from the west it would make the return trip a down wind sail.

The Vikings rationally judged chances of success or failure. They were expert seamen, and they knew they had a boat they could trust. They would usually not take on a task that was not feasible. In raiding, they had the option of hit and run as they had the superior craft. At sea, they would go as far as they liked. Common Viking sense was applied in navigating.

They would set out into the ocean in the hopes of obtaining something, visiting someone or finding resources that could help support their barren life in the cold and rocky home of Norway. With a good boat, sufficient provisions and warm clothing, they sat sail out into the cold and empty sea knowing that they would find land if the wind and weather would cooperate. And they always had the option of turning back. To the east, Norway would be waiting. Even if they veer far off-course, they could not miss that target (Engvig, 2001).

THE BOAT'S ORIGIN

Our Viking boat was very old, but in original and seaworthy condition. It was not a Viking replica, but an original open Norwegian longboat built, down to details, the same way as a Viking boat was built more than a thousand years ago. A famous boat builder, known to make superior boats, built it in 1863.

Every one of the many iron rivets was keeping up well. The boat survived all these years because it was an excellent vessel. When the original owner discovered that he had a boat with better sailing ability than all the other boats of the fishing fleet, he saved and protected his boat. He only took it out of the shed on special occasions. Otherwise, it could stay inside for years at a time. No one could overtake his boat when he sailed it. This was an important status marker along the Norwegian coast. Like the farmer that had the swiftest horse, it was prestigious to have the fastest boat.

He got himself another vessel of the same size for day-to-day usage, one which he completely wore out while the old one was kept high and dry in the boat shed. His grandson only used the old "firing" for five years during the winter cod fishing season when he was young. He wanted me to have the boat and use it for research purposes and to show the world how to sail square sail. I named the boat *Hitra* after the island where this old man lived.

Firing Hitra was built in Aafjord on the Fosen Peninsula more than a hundred and forty years ago and is still all original. It has never been restored. It is a genuine open longboat built like the Viking longboats we know today. *Hitra* is close to 30 feet long, 6 feet wide and draws two feet of water when fitted out. It carries one mast and a single square sail of 14 square meters with several reefs. It also carries 10 oars. It is made from slow-growing Norwegian spruce from the coastal forests at Aafjord. The bottom boards, well over 12 inches wide, must have come from a tree that was young more than 300 years ago. The boards are about 5/8 inch thick and are "sawn" or riveted together with *saum* (iron rivets) and *roer* (roes) made locally at the boat builder's smithy.

All the twisted bands in the framing are roots or branches of spruce or pine. They look delicate or under-dimensioned to an inexperienced viewer. Needless to say, every part of the boat is of matching strength and will keep together as long as the people aboard have the knowledge and ability needed to handle this fine vessel the right way. The whole boat is extremely flexible. It only weighs 500 pound, and needs quite a bit of stone ballast for sailing.

It can be rowed at a speed of five knots and travels by sail at 10 to 12 knots over time, but can reach 20 knots or more during surfing conditions. *Hitra* has sailed from Horten to Oslo in four hours, which is the same time the local steamer would take, and logged more than 50 nautical miles along the shores of Great Britain as in Norway in less than four hours with little or no current. It was common for us to outrun modern yachts going downwind.

This famous painting from Oscarshall depicts a Viking ship. It is from the same period as when the *Hitra* was being built. At that time academic knowledge on the design of Viking ships differed a lot from that of the common boat building design. Later Viking ship excavations revealed the truth. When it was excavated in 1880, many refused to believe the Gogstad ship to be a Viking ship.

The important thing to keep in mind is that the *Hitra* was constructed more than 20 years prior to the very first relatively complete Viking ship was discovered and excavated. The boat's builder in Aafjord, Norway would have had no knowledge of how the Vikings constructed their vessels. In 1863, the general belief was that a Viking ship was a sophisticated vessel with a lot of bells and whistles that make them even more impressive, much like the way Viking helmets were adorned with horns and accessories. Paintings depict this common opinion. Anecdotal evidence indicates that many refused to believe that the Gogstad Viking ship was a real Viking ship. When excavated in 1880 as the first relatively complete and thousand-year-old hull, it was too much of a "copy" of the open longboats still in use on the coast of Norway where *Hitra* was built. Was someone playing the scientists a trick by burying an old boat a hundred years before? Also it was too well-preserved to be that old. For the archaeologists, this was not an issue. There were many genuine Viking artifacts and valuables found along with the ship. This was "the real thing" and the world finally learned how Viking ships were constructed. Our old workboat *Hitra* was built in the same way with the same details. It is the missing link and represents the genuine alternative to all modern day copies of different Viking ships.

Contrary to the ship itself, no really informative samples of Viking rig, rigging and sail has been found. Modern replicas of Viking ships are usually equipped with an adjusted 19th Century rig, adapted from the one that is the original in my boat. We believe that rather than trying to conduct this experiment with the use of a modern replica, we would do it simple and use this open longboat just like an average open Viking boat would have been used. We would not equip our boat with new gadgets that so many Viking voyagers do. Some Viking enterprises even include a warm and dry shelter on board with radio and stereo music, a galley for hot meals and dry underwear and a dry bunk. Not to mention the diesel engine for propulsion and bailing and electricity for all the navigational aid needed. Nothing is farther away from the sailing experience of the Vikings.

The very first full-sized copy of a Viking ship was made when our *Hitra* was 30 years old. The replica was to be presented to the world at the 1893 World Exposition in Chicago where copies of Columbus' ships were also present. They were not sailed to America. Kaptein Magnus Andersen decided to sail his ship all the way to Chicago, and so he did (Andersen, 1895). Not only that, of all the expeditions done in a replica Viking ship, I consider that first voyage to be one of the most scientific Viking sails ever done. The ship was named *Viking,* a copy of the first excavated Viking burial ship from Gogstad, Sandefjord, built by nearby Framnaes Shipyard. *Viking* was sailed by professional seamen that had practical experience in sailing the high seas in Norwegian windjammers. They were used to hash conditions and foul weather out in the ocean. Captain Anderson did use some adaptations and modern navigational equipment. However, he stated that a true Viking sail should be done without these alterations. *Viking* was constructed using the same lap-streak techniques as in all traditional Norwegian longboats.

TESTING THEORIES

We wanted to pick up where Captain Andersen left; go back to the roots as far as we could, start from there and make it simple. I was convinced that the *Hitra* could do the boat's part—the job of bringing us across the ocean without the old iron rivets coming apart. In my opinion, she would keep together, even on a long voyage. There was nothing very spectacular about our sailing, but it did get some coverage from local media, which showed support for our little group and was interested in our daring project to test old iron and sail without any modern navigational aid. Also, we had done previous long distance sailing without modern gadgets, except for life saving equipment, so I felt confident that the sea and the sky would guide me as it guided people in the old days when they traveled out to sea without support equipment. The old navigators depended solely on their knowledge, experience and intuition to find their way under various conditions. They only feared storm-induced waves that could damage the boat, and getting lost in foggy conditions.

We, of course, had an extra dimension to focus on. We were also testing out the durability of 19th Century craftsmanship, since we were going in an historic boat that had never been restored or renewed. *Hitra* was the exact same boat the boat builder had put together in 1863. We even had the original mast, yard, sail and other parts. We did make a copy of the sail, renewed almost all the hemp lines, repaired some of the oarlocks and made several new oars. From my lifelong experience with traditional Norwegian longboats of this Aafjord type, I expected the boat to keep up if I did my part right and let the old boat have its own way if needed in running away. Although these crafts were built with light wood, the final construction is surprisingly strong and flexible when all the pieces are skillfully riveted together. When the ship cruised, you can well see how the bow negotiates a big wave by flexing. This ancient design was developed with knowledge and respect for the ocean, to work in-line with the wind and the sea, and not to fight it. If the *hoevedsmann* remembers this, there should be no problems negotiating between six-feet high waves without.

Our claim is that this way of sailing is a more appropriate approach to obtaining information about the ancient ship design's durability and how the Viking did their sailing and navigation. It took me a long time, almost my entire life, to educate, prepare and motivate myself for this endeavor. My father, Thormod Engvig, was a professional navigator born in 1886 and raised by a Norwegian fjord. He was a seaman whose life started out in the fast iron built British full rigger the *Zealandia*. He subsequently moved on to coastal schooners in America, got officers- and masters papers and sailed in troop transporters, Lakers and tankers before finally ending up as a master, in command of American steam tankers such as the *Huguenot*. He taught me to tell the time just by looking at the sun. He had a fine Howard watch, but never used it at home. He did not need it because he always knew exactly what time of day it was. He was always at home when I grew up, as the Germans kept him in house arrest during World War II. During this time, he taught me how to behave in a small boat, to tell time by looking at the sun, and to always know directions ("to keep a compass inside your head") on sea and land, and to feel confident in a small open boat at sea.

Then I learned square sailing from Jakob Kvithyll, better known as "The Last of the Vikings." He was an old man living in my community when I grew up, and I was his only student. All this happened before I went to the cadet ship at the age of 15 and continued on to a tanker as a seaman. I worked my way around the world before I was 17, crossed the equator 12 times, and I sailed in a motor ship, a steam ship and a windjammer while still a teenager. Later, during my sociology studies at the University of Oslo, I interviewed seafarers and boat builders on the coast for the Norwegian Maritime Museum. Old fishermen told me they never had a compass or carried a watch in their boat when they went out into the ocean. They used the sun to tell the time and direction, and with the prevailing wind from the west, they felt confident they could find their way home.

Høvedsmann Jakob Kvithyll, born 1884

Captain Thormod Engvig, born 1886

Reidar Melkvik boat owner, born 1907

During my many conversations with old boat builders, I learned that they often worked in pairs when building a boat. In order not to be in the way of each other, they would each work on one side of the vessel. This could sometimes trigger a fierce competition to see who finished first. When educating a young boy in this skill, the lad would usually first hang around and assist the builders. Then he would split wood in the forest, cut roes or make ship-nails and rivets in the smithy. These were all preliminary work that had to be done. Even tool making was part of the education. When the youngster was ready to make his first boat, the grandfather would usually be the teacher and they would start with a small and simple boat, like the smallest *faering*, making "one side each." To make a boat that could be sold, the two sides had to match, so the youngster had to copy the old man in every detail. This is a smart and simple way of preserving good knowledge and skill. Boat builders at Aafjord in the middle part of Norway constructed the same boats for more than a thousand years. Basically as long as the boats were in demand, the builders knew, as did the users, that any alteration of the boats' design would detract from its usefulness. They always worked during the daylight hours, and in Aafjord during late Spring, they could start well before four in the morning and often did. However, they were never allowed to start riveting before six o'clock in the morning as the women and babies on the farm needed the sleep. Riveting always made a lot of noise.

Strangely, many of the best boat building farms that had large forests of particularly good wood also made good iron, and in the buyers opinion, provided the best raw materials and did the best ship building job had little or no knowledge of the sea. These farms had no access to the ocean and were located many miles inland from the bottom of the fjord. They had little or no use for even a *faering* themselves and did not keep a boat for their own use. If they needed transport by sea, they would depend on the locals for assistance. This also tells us that a man who wanted his boat to be rigged had to get a rigger, who usually lived by the sea, to finish off that part of the job. A sailmaker was needed in getting a new sail. That is, if the buyer did not do all this work himself.

The introduction of oil engines around the turn of the 20th Century led to the demise of traditional open longboats. The engines were shaking so badly that much stronger hulls were required for support. The lightly-built open boats which used oars and sails did not last long after they got engines installed. The new generation of sea vessel was the sturdy motorboat or motor-cutter. The demand for new open longboats gradually disappeared.

On our first voyage in the ancient craft we discovered to our surprise that we could outrun almost any modern sail and motor vessel with the wind abaft. We finally decided to offer a tow.

During rain in summer and for rest and sleep, we would find shelter under the sail, but discovered that the open boat was a wet place in general, especially when sailing well. Keeping dry was always a challenge.

WESTWARD

After several years of sailing my own boat *Hitra,* I knew her speed and performance under different wind and weather conditions. This is very important if you are to estimate your position at sea. I finally decided I had the theoretical and practical background and the knowledge needed to sail "The Viking Way," a thousand years after the real Vikings. David Lewis' books on traditional navigation in the South Sea inspired me. To sail straight out into the North Atlantic Ocean with absolutely no navigational instruments or aid whatsoever also required more Viking studies, and practical work on all aspects of the venture. Interestingly, the saga usually measures distance by numbering days of sailing. I also needed time for planning and preparations, including talking people into going with me on this strange scientific mission.

Most importantly, I needed two boats. That was my biggest worry. The following vessel should, of course, be of the same type, design and rigging, to reenact Vikings going abroad. Vikings often sailed together in several ships. That meant the follower could not be a modern boat with an engine and a film-crew.

I had a candidate, a beautiful creation of a longboat, which had been built by the same boat builder as the *Hitra*. It was said to be his final masterpiece. This boat was stored in my boat shed in Rissa. But sadly, it was not seaworthy. That was the follower I would have liked to use. It was built in 1872, only nine years after the *Hitra*. It was two feet shorter, and it was a most daring construction, the finest boat ever made in the district, according to legend. But it had a bad reputation. The owner was advised not to use it without enough ballast. He went sailing one day in strong wind with his young son. The boat flew across the waves like an arrow, and then it capsized. The old man drowned, but his son hung on to the keel and was saved. This happened more than one hundred years ago. The son and his neighbors put the boat back in the shed and left it there. He never took it out again. It was used as storage for old fishing nets. His son was an older man when the roof of the shed caved in during one stormy winter. No one cared to do anything with the roof or the boat in the old collapsed shed.

When I learned about the story around 1970, I went to the shed and found a beautiful boat. It was 27 feet long, looked like a gondola and had 8 oars. It had absolutely gorgeous lines and appeared as if it had not been used. But the bottom was quite rotten, although the keel looked good. I bought the wreck, took all the old and rotten nets out of it and discovered that the boat was in a very bad state of decay and definitely not seaworthy, but I still wanted to save this gorgeous, sharp-looking creation if it was possible.

In the local store I found a roll of sturdy construction plastic. I got the plastic under the keel and eased it up the sides and nailed it down with long thin pieces of wood on each side. I carefully launched the boat and it stayed dry! It was strange to see kelp and the bottom of the fjord right under the keel when I got on board. My daughter Gunn, who was six years old at that time, and I took the longboat under tow and moved her 18 nautical miles to my home where it went inside my boathouse.

Later, I shipped it over land to my local boat restorer, Roald Olden at Aafjord. He marveled at the beautiful boat, but refused to restore it. He said: "It is too far gone. It is not possible. I'll copy it. I can make you a brand new boat, using all the measures and particulars of the old one. I will also take off the good pieces and use them on the new boat." There was basically no way the old boat could be saved. It was falling apart from rot and decay right before our eyes. An exact copy of the boat was made during that winter. The boat builder did a very fine job copying the old original from 1872, but only two small genuine parts from the old boat were good enough to go on the new one. I named her *Frøya*, and the boat was rigged and made ready to go as the following ship for the *Hitra*, with me responsible for finding our way across the open ocean as *hoevedsmann* on the *Hitra* when we were ready to do that test.

Roald Olden the boat builder (right) shook his head. This fine boat from 1872 was not in a restorable condition.

"I will make you a new boat, an exact copy." Olden used all the measurements from the old boat in designing the new one.

PRELIMINARY TESTS

It started when I sailed the *Hitra* on a coastal voyage of more than 600 nautical miles along the coast of Norway from the island of Hitra by the Trondheimsfjord to Oslo without much more aid than the Norwegian Automobile Association's road book. We depended on the old knowledge that the Vikings had: Norway would be to port all the way. We followed the coastline and our major means of establishing position were the many ferry paths we had to cross. They were outlined in the road book. This first long haul taught me what an amazing vessel I had. For most of this voyage, the *Hitra* flew along the coast under her sole square sail with a fine breeze from the North all the way to the Ness or Lindesnes, as the South tip of Norway is called today. As we rounded the headland, the wind sprang up from the west and took us straight into Oslo in 12 days and four hours. The voyage was mostly sailing by day only and with several significant stops because of little or unfavorable wind, sightseeing and informing locals on our mission.

Keeping the small logbook dry was difficult. Protecting gear from wet summer rain and spray from the sea was a major challenge. To bring along a whole stack of sea maps for navigational purposes would not have worked. Maps are expensive; they would most likely be ruined. But most of all, we really did not need any, just like the people who sailed the same way hundreds of years before us. We learned how we could use the sail as a tent for a night's stopover and some hours of sleep in the boat.

The boat did great and had no leaks, proving that the old iron rivets were capable of keeping the streaks together. We had few problems with strong wind and high seas. But we experienced a couple of close calls on Nord-Vestlandet with high mountains and narrow fjords. The conditions created strong gusts of changing winds coming down from the mountains, blowing the sail aback, or kicking in so hard that it would have capsized the "Hitra" quickly if we were not aware of what was coming and let go at that very moment it kicked in. But we were prepared, and saw it coming. The secret is to let go the sheet and the halyard and reach for the down-hauling line. The sail will take off and behave like a wild horse but most of the power pulling the wrong way will be released or diverted, preventing us from rolling over. On the coast of Trondelag, further to the north where the boat came from, the land is lower, and the wind tends to stay stable, blowing from one direction like on the ocean. This is why *Hitra's* full sail is too big to carry the boat under difficult conditions and very strong wind. With our reefs and clews, we could decrease the sail gradually and at the same time descend the sail on the mast until it became a small studying sail just above the gunwales for down wind travel in a full North Atlantic storm.

Sailing north of Bergen was a lonely experience. We hardly saw any other sailing crafts, but encountered quite a bit of commercial costal traffic and had to watch out for car ferries, as they tend not to give way to anyone. Small coastal freighters had a hard time overtaking us when the wind conditions were in our favor. This was a new lesson. The old *firing* was a speedy vessel running in the wind.

Our greatest surprise was our performance compared with modern crafts along the coast when we got further south to Sorlandet. Summer traffic was busy in some areas. To our great delight, we showed fishing boats, modern yachtsmen and other crafts that we had the speedier vessel. Again and again, we left other boats behind when the wind gave us a good push. It got to the point where I was offering to tow modern yachts as we speeded past their bow. We were in the oldest boat afloat and no one could catch up with us.

Under way on the Viking coast for the first time. *Hitra* is leaving land at Hustavika.

AROUND THE NORTH SEA

To show the world how amazing this old boat design was by doing a greater deed became my new objective. It would be a grand experience if this old, flexible and delicate boat made of wood and iron would keep up. My private notion was that it would. When Magnus Magmusson and a team of British scholars came to Norway asking for support for the archaeological excavation of the Viking site at Coppergate in York, England, I was ready to offer my support and embark on a new challenge. It was a site of Norwegians emigrants they were excavating. We were told that some 10,000 Vikings were living in York at that time. They really left their signature on the city of Jorvik, today's York, and the countryside of Yorkshire that still is full of Norse farm names. The Vikings ruled the old city of Jorvik even before they developed cities in Norway. They rearranged the city's outlook by making new lot arrangements and staked out new streets. They were given Norse names that ended in "*gate,*" which means "street" in Norwegian even today. My daughter Gunn and I became the Norwegian voluntaries. She was 13 years old. We were more than happy to offer our support during a Viking sail to England.

The York Archaeological Trust provided us with a stack of promotional pamphlets that we presented to the media in cities around the North Sea on our first mission to York. Our goal was to keep land to the left all the time and ask locals where to go.

Scientific expeditions can be like a summer vacation on fine days and remote areas like this on the west coast of Sweden.

Gunn, only thirteen, in command of the *Hitra* under a full sail hugging the shoreline off the coast of Belgium.

On this first international mission, I would only be using the original historic longboat *Hitra* and the Viking merchant Ottar's description of how to get there. The English King's scribe wrote instructions more than a thousand years ago. This record provided a simple and useful recommendation we intended to use. Basically it states: keep land to the port all the time and do not go in where a lot of water goes into the land. That would obviously be the Baltic Sea. So we sailed around the North Sea according to Viking Ottar's description.

Our verdict is that Ottar was right. It was quite simple. Even if I knew the basic outlines of the North Sea by heart, the key was to hug the coast until the White Cliffs of Dover could be seen far away in the distance. That would be to the starboard. To get the land to port we then had to turn the *Hitra* north again. From then on, we had England to port as we had Norway, Sweden, Denmark, Germany, Holland, Belgium and France to port before we got that far. The goal of the mission was to promote the Viking excavations at Coppergate, as well as testing the boat's construction and my navigational skills on a long haul. All the major media offices in the countries mentioned were visited by the young crew from the very old boat, who handed out information and asked for support. My only crewmember was my daughter Gunn. The story became one of the highlights in Northern Europe that summer, and the York Archaeological Trust was granted another year from the developers to finish the excavations at Coppergate.

We used 38 days on that voyage, stopping over for a day or two in places like Copenhagen, Amsterdam, Breskens and London. The greatest challenge was to get safely across the English Channel. We were almost hit by enormous merchant ships speeding out to sea, big ferries and hovercrafts. The Vikings would not have had that problem! The only time we were confused about the route was when we headed for the river Medway instead of the Thames going up to London. By chance I shouted to a small local fishing vessel and asked for direction. He pointed us to the right course.

Our greatest pride was when we sailed up to the dock at Greenwich where the world-famous clipper ship the *Cutty Sark* is berthed in a dry-dock. She is a full rigged square sail ship, with her sail and rigging based on the same principles as in the *Hitra*, or in any Viking ship with a square sail. Any square sail rigged ship from the age of white sails is sailed, handled and reacts in the same way as our *Hitra*. All the basics are the same even if the big windjammers usually have three or four masts behind each other, and often two masts on top of the lower one. They have split the original sail into four, five, six or sometimes even seven sails on top of each other to be able to manage each one of them. Most of all, these two proud and speedy square sail vessels were built in the same decade and their construction material was wood and iron. The *Hitra* was built in 1863 and the *Cutty Sark* was constructed in 1869. The latter has not been to sea for many years. To have an original and complete square sail rigged boat sailing all the way from Norway to visit a contemporary surely qualifies for an entry in the Guinness Book of Records!

Meeting of two aristocrats. The *Cutty Sark* docked at Greenwich gets a visit from another square sailing vessel, the *Hitra*, built in the same decade as the famous clipper. They are both made from wood and iron.

WOOD AND IRON

Like the *Hitra*, *Cutty Sark* has a wooden hull, but one that was built on an iron frame. Both were built in the same decade as the two other principal iron ships described in this book. The *Star of India* in California and the *Hansteen* in Norway. The 1860s was really the time when iron influenced ship building greatly. However, not everyone was willing to abandon wood entirely and compromises were made. The *Cutty Sark* is such a ship. She is a composite ship, and to my knowledge the best preserved ship of this construction anywhere in the world. Her frames and other important parts that require extra strength are made of iron, while the outside hull is made of wood.

In the *Hitra's* rigging, only the four fastenings for the rudder are made of wrought iron, and the wooden deadeyes for the shrouds are secured with iron. But in the boat itself, the only iron parts are the nails and rivets that hold the outside hull together, just like in a Viking ship. The rest of the fastenings in the *Hitra* are all made of wood, the primary ones being those that goes through the hull and hold the bands or frames to the inside of the boat. All the bands are secured with wooden pegs. On our way to York we encountered unusually short and punishing waves on the shallow waters of the Ijsselmeer. One of the wooden pegs in the bow section of the boat broke and the boat started leaking badly. We had to bail water almost constantly while we tried to mend the leak at sea but had little success. So we sailed the boats up on a mud bank when we reached shore, moving all the heavy stuff aft before landing. We emptied the bow and with the traditional tools we brought along, got the broken peg out, and made a new one

Repairs of a bad leak due to a broken peg are done by drilling it out and replacing it with a new one made on-site. Most damages like this can be dealt with temporarily while at sea and repaired at the first landfall.

from spare wood kept on board. We drove it in and secured it on the inside. This emergency repair job later held up during long voyages across the open ocean. Never did any of all the old iron rivets give in. Even now, they are still intact in the boat, and that is the reason why the laboratories have not received a sample for analysis.

We did learn something about iron rivets and the breaking up of a longboat during one of our later expeditions to York. At that time, I had gathered an "invasion fleet" of five open Norwegian longboats with square sails at Hull for a final Viking "attack" on York, more than a thousand years after the heydays when Norwegian Erik Bloodax was the King of York. This fleet consisted of the two boats *Hitra* and *Frøya*, (both of which had sailed about 700 nautical miles of open water across the North Sea from Norway), the *seksring Rein* from 1905, the *big-faering Rissa* built in the 1890s, and the *small-faering Ormen*, built in the 1930s. The latter three had all been shipped to Hull on a freighter. We assume, however, that the Vikings would use somewhat larger vessels than the rather small inshore crafts we used.

This flotilla of five open longboats left Hull and sailed under the new Humber Bridge and picked up the inrushing current after we had entered the River Ouse. I knew the tidal wave was coming on our tail with full force and gave all the others the advice to be on the outlook and keep to the middle of the river, as it would be a roller coaster ride. I had the *Rissa* in tow after the *Hitra* and when we got to Selby, we got into trouble. The towing went astray in the strong whirling current and the *Rissa* ended up alongside the *Hitra*, thus limiting our steering ability. We were approaching a large concrete bridge pillar at full speed and the *Rissa* hit it first, taking much of the blow, while the larger *Hitra* suffered minor damages. The *Rissa* was completely crushed against the bridge pillar and finally disappeared under water. I cut the towing line, continued through the town and caught up with the rest of the fleet. We waited for a while but saw no traces of the *Rissa*, and came to realize the faering was a complete loss. We conducted the "invasion" with the four remaining boats. The fleet had no problem "conquering" York. The Lord Mayor was at the dock and was the first to "surrender." "Blue Peter," a well-known BBC TV show for teens, made a fine program on our mission to York. Even Japanese TV was there.

The year after, I was back in York for the Viking regatta with the faering *Mr. Thomas* built in the 1870s. A few parts of the *Rissa* had been recovered from the river and brought to York for display at an archeological exhibition, but surprisingly little of the ship had been found even if the word was passed along the river to look out for debris or parts of a 20 foot long lap-streak longboat. These last remains of the *Rissa* told an important story. It was a part of the sides about 10 feet long and with all the riveted lap-streak connections intact. The boat had split open about half-way between two strikes of rivets. I had seen the same remains of damaged boats several times before and it strongly indicates that the riveted areas are the strongest part of any longboat. I have never seen rivets pop out, or a riveted streak split open because the old rivets had given way. The area with iron rivets proves to be stronger than the rest of the boat even in a boat that has been used over a very long time. The conclusion is that as long as there is a fair amount of iron left in the rivets, this will be the case.

In riveted ships with plates made of iron, it is more common to see the rivets break off from the plates if extreme forces have been applied. Photos from a gas explosion in an oil tanker, a collision between two big ships, or a boiler blow-up in a steamship tend to show us twisted plates with series of holes after the rivets popped (Engvig, 1984).

The voyage was very exhausting at times. The young "Viking" girl Gunn is sleeping on her off-watch with the lapstreak boards as the pillow. Notice the old rivets with their home made roes along the streaks.

"THE NICE VIKING"

It has been more than 1,200 years since the Vikings landed at Lindisfarne and raided the monastery on that Holy Island. Today, this incident stands as an overture to the classic Viking Age — an era of raiding, pillaging and plundering. During our 38 days of sail around the North Sea in the *Hitra*, Gunn and I had time to think more about how the Vikings got to the British Isles. It is believed that

Yorkshire, the City of York and other areas on the British Isles and in Ireland saw many thousand Viking settlers. They made a significant impact on York, from crafts and trade to administration and redevelopment of the whole city structure. Yorkshire's countryside still has many names of Norse or Viking origin originating from farms or small settlements of people that moved from east of the sea. These friendly invaders, known as immigrants today, helped develop the land and left lasting marks along this part of the East Coast of England. They were "The Nice Vikings" who were never mentioned in the Saga. It was the bad guys and major dramatic events that got recorded and became part of history.

A greater number of the thousands of Vikings that moved out of Norway during the Viking Age were gentle and well-behaved, and never pillaged or plundered. They left for the new land alone, in families or groups to get away from famine, extortion or poverty. They brought their skills and artisan knowledge along with their culture and their most important belongings, like craftsmanship tools. Some of the items that we brought along on our expeditions for the use on board the boats were exactly the same tools as the Vikings had brought, according to the finds from the excavations. We had no prior knowledge of this until we came to York on our sailing mission. The best example was the hone we had on board the *Hitra* for sharpening of iron knives and other sharp artifacts of iron. Ours came from Eidsborg in Telemark and might be very old. The York Archaeological Trust has established that some Viking Age whetstone excavated at York also came from Eidsborg in Norway (Tweddle, 1986). They were exported during the Viking Age and later, and the Coppergate dig confirmed this by unearthing several Viking whetstones (Hall, 1994).

People at the YAT fully agreed that the majority of the Vikings were nice and friendly people. We were presented copies of items from the Coppergate site on the site by chief excavator Richard Hall. They were wooden cups and bowls the Vikings made for eating and drinking more than a thousand years ago.

I believe many emigrating Vikings would have used their small open boats when they headed to England. We have shown that this is not very difficult, even though it is a long voyage going around the North Sea. Our main argument for this theory is that in the Viking Age, Norwegians were much more dependent on their boats than today. They were living by the fjord, bay, *vik* or on the island. Travel over land was difficult and exhausting, and it was limited what a person could bring along.

The fjord was the Vikings' main highway, even inland lakes and rivers served the same purpose. Up until a hundred years ago, communities across the fjord had more interaction and stayed closer than the neighbors that lived nearer in actual distance when people had to sail or row all the way around the *nes*, or walk on foot across the mountain ridge. Access across many ridges full of ravines, boulders and other natural obstacles would be impossible for a man and his horse. Today this is quite the opposite. There is little or no contact across the fjord any more. Contact and interaction are concentrated along the many roads around, or or through the mountain.

The *faering* was the common means of transportation during the Viking Age. It was needed for communication, fishing, egg collecting, shuttling of wood and many other functions. Small boats are easy to make and inexpensive compared to a large Viking ship. There would be relatively few larger ships built compared to smaller vessels. Like today, an average citizen could own a small boat. But modern people would rather depend on their automobile than an open boat for speedy transport. Many Viking Age men never got on board a big Viking ship. A good inshore vessel would be sufficient to move a family or some adventurous youngsters like us to Normandy or England, where they established a new life, made friends and started families.

In an open boat, everything tends to get wet. Bread becomes moldy. The normal procedure is to cut away all the smelly green stuff on the outside. It is fed to the fishes. The loaves are then served to skipper and crew with lots of "goodies" on them. Hungry as we were, the food went down without a remark.

OCEAN NAVIGATION

There is always the long road, and the short-cut. Cutting straight across the ocean is quite another challenge in comparison to coastal sailing. My boats were of the larger inshore types. Today they are considered rather small to use on the high seas, but we do not know the smallest size for an ocean going vessel during the Viking Age. Any well-built boat between 25 and 30 feet would be sufficient to take a small party across the sea to England, Ireland, the islands in the Atlantic, or even further, and back. Consequently my boat's overall length or small freeboard did not stop me from trying to apply my sailing and migration theories and test my navigational skills by attempting ocean sailing. I had acquired a lot of experience with my boats during the previous voyages and felt that the vessel was good enough to stand this test. It would be a sweet victory to take what we believe to be, the oldest complete and known seaworthy vessel of this traditional Norwegian longboat type straight across to the other side only by sail. Shetland is the nearest land, which lays about 200 nautical miles to the west of Bergen. But these islands are small and without high mountains. If we miss them, the next land mass would be America some 3,000 miles away.

Traveling west into the setting sun is a stereotypical but symbolic scene that had been reproduced in many Western cowboy films. To us, it was far from any mythic notion. We took the bold challenge to do what no one has tried in modern times, to find land without any navigational aid whatsoever. No maps, compass or watch was carried on board. We were determined that the sun would show us the way. And, indeed, it did.

The Challenge

Our challenge was to attempt something that no one in modern times had tried before: to sail straight out into the vast North Atlantic Ocean without any navigational instrument. Furthermore, we only carried a lone square sail and the 10 oars that we were too short-handed to handle. We depended on the wind to take us across and the sun to guide us, and we were confident we would find land.

In mid-June, the two longboats *Hitra* and *Frøya*, with six crew members, Gunn, Stig and Marit, in the *Frøya*, with Knut, Inger and I in the *Hitra*, were ready to go west from an island north of Bergen. Safety gear and equipment, provision, other stores and spare clothing were secured in sea bags. Two inflatable life rafts, two set of emergency equipment, six Helly-Hansen survival suits and an emergency beacon were the most modern gadgets we brought along. We did bring a brand new position indicator and a battery, which was placed in the following boat and coordinates would be taken in the mornings and evenings to later compare with my logbook and see what the error might be. We waited for two days in the rain and fog before the wind turned.

On a Saturday afternoon, a breeze from the northeast picked up and we decided to get underway. A short row brought us past the tip of the headland where we hoisted sail and the course is set straight into the vast Atlantic Ocean. We were sailing with the sun about 30 degrees to port of our beam. In two hours, it was more in our bow and I estimated the time to be 6 pm. The wind was nice and steady and we were going at a good speed, steering straight west into the sun. It was close to mid-summer and by approximately 8 pm, Norway was falling behind far away aft and the sun was on our starboard side 10–15 degrees above the horizon, much as I had hoped for. We carry a full sail and have the head of the sail all the way forward on starboard. All the three members of the *Hitra* crew were sitting close to the gunwale. This was spectacular sailing into the sunset. The speed was about 6 knots. The boats ran nicely and gently, but took a few splashes on starboard bow every now and then. This is the well-known signature of the low and speedy open longboat.

The *Frøya* followed behind like a shadow as the sun set into the ocean. Some forward cloudbanks obscured the view. The sky was beautiful, red and gold to the north deeper blue the other way. *Frøya* had a golden vane in her masthead. It was glowing like a lighthouse in the twilight. I was impressed with its signaling effect and I challenge the theory that they were used as weather vanes on a Viking ship. Even in strong wind, they would continue to move and spin around as the boat cut through the sea. Even the noise they make became a nuisance. My later report concluded that they would be much more useful for signaling and identification close to land (Engvig, 2003).

Heavy coastal traffic of big tankers and cargo ships that we saw a while ago were all gone. As we approached midnight, the moon was up aft on port, the wind was steady and the speed was good. The "white night" was peaceful and light.

One or two faint stars were visible; one could be a planet. Gunn was doing very well in *Frøya* while Stig and Marit were off-watch and asleep. I was ahead in the *Hitra*, with Knut and Inger in my boat also trying to get some sleep.

The wind was getting stronger from the north and we were picking up speed. We were both sailing as hard as a full sail allowed and I estimated the speed to be close to 10 knots. If that continued, we would reach Shetland in two days. Sun rise was not registered. The sky to the north and east was overcast, but we were sailing as close-hauled as we could, trying to keep the bow up and towards the west. I tried to gain some of what I have lost. I felt that the wind was setting us towards the southern part of the Shetland Islands. Suddenly, the sun broke through the clouds. It was about five degrees above horizon and four notches aft on starboard. We were still heading west. But the wind was my determining factor. I could only do my best in keeping the head of my ship up towards the west, and hope that the wind would take us there. All ships with square sails, from my open boat to the windjammers, are dependent on this.

Dawn came early on our first day at sea. It was easy to follow the rising sun. By only using my hand and a pencil, I could tell when it stopped rising and when it started to descend again. I was confident I could determine mid-day with a half an hour margin. The sun has started to descend and we are getting closer to Shetland by each hour as *Hitra's* bow was pointing a bit north of west. The sun was still aft of mid-ship. That was promising. I felt confident like the old fishermen whom I talked to, knowing their way at sea was as natural as walking. If this keeps up, we would probably reach Shetland sometime the next day — we were half-way after only 24 hours of sailing. But we were in for a change. The horizon forward to the northwest was pitch black, and it was growing on us.

Frøya leaving the coast in pursuit of the *Hitra* heading west.

The weather changed mid-way through the voyage. We stayed close together. Strong cold winds from the north came down on us. We were forced to reef down and put on survival suits. On the top of a wave, we were riding many feet higher than the *Frøya*. We could look down into the bottom of the boat way below us.

Changing Winds

Bad weather was coming our way. So far, we only had simple meals consisting of bread and water. We had oilskins on, but I ordered all members to put on survival gear. Mine was drenched by a large wave as I was putting it on. Knut also got wet. We reefed down our sail to be prepared. Things looked bad. But the feared squall had no wind, only a little rain. The waves that followed were rather huge — more than six feet in height. We had been falling off to the south to be prepared for a wild ride, but were soon back on course. Unfortunately, our speed dropped to only a few knots.

Suddenly the wind kicked in and the waves intensified. With a reefed-down sail, we were shooting out of waves down into a valley and up the next one. When *Frøya* was between two waves, and we were on top of another. We look down at *Frøya* as if it was on the ground floor while we were looking at it from the second story of a building — it was surreal. We were sailing close together. The next moment, we could only see the very tip of their mast; the rest of the boat was obscured by the big wave between us. The experience was unbelievable. Suddenly, more than half the boat was in the open air when it shot out over the top of the wave, gently tipped forward and rushed down the surfside. That was spectacular sailing. We were doing good speed again with a reefed-down sail less than half-up the mast. It was fun, and I regretted not bringing a film team and a watertight camera to capture our adventure on film. We were playing hide-and-seek with the big seas and felt confident we had the upper hand. The old iron rivets were keeping up very well.

With our partner few yards away, we could only see the top of the mast and sail when a big wave came between us. Orders were to stay together and watch for breaking waves. After dark, it became a real challenge to keep contact, once we had to use flares to re-establish contact. We lost the battle to reach the Shetlands.

For a while, I did not really care if we did not end up where we planned to go. We must be heading to the south or southwest in the evening. Big black clouds made an early and dark night. The wind was still strong and the waves were big. It was raining. The two boats had problems keeping contact. Our kerosene lantern's light was easily lost when the boats were in between waves or when the sail obscured our sight. That night was a nightmare. We lost track of each other several times. At one time, we had to use flares and that helped. We finally decided to call it quits. We lowered our sails and tied the boats together with a 300 feet long line. It was probably pass midnight. *Hitra* carried a sea anchor and it provided a perfect solution. We were in the middle of nowhere out in the sea. We went to sleep, with one person keeping a relaxed watch. I got my first two hours of sleep in my survival suit. The storm had raged on for more than a day and it was cold.

The new morning was gray. Everything was gray. The sea was relatively calm, but it was wet everywhere in our boats. We were all freezing cold. Everyone's teeth were chattering uncontrollably. We started sailing. The wind was determining our course. It was coming from northwest as far as I could tell. After a while, we picked up good speed but were forced south, and I told my crew that we were east of the Orkneys at the moment. What I did not tell them was that I was afraid we might end up in Holland, way to the south, if we kept on the way we were heading. It was just a worrying possibility that I kept to myself. I had no real idea of how badly corroded *Hitra's* old rivets might be. What if the iron rivets gave in and the boat split open? Fortunately, there were no signs that this was going to happen. I trusted that the old vessel would stand the test as I knew we did not force it to work against the elements. The wind and the waves were the determining factor out here, so I suppressed the creeping thoughts and hoped for a new wind that would take us west instead of south.

We had great difficulties preparing food. The bread was all soaked in salt-water before butter could be applied. We brought 30 loaves of bread and were still working on the first one on board the *Hitra*. Nuts, raisins and sweet biscuits and water from our oak barrel made up our daily ration. It was impossible to change clothes. Trapped seawater and sweat created a most curious odor, which escaped from the neckline of our survival suits. Our spare clothing was wet as well.

The waves splashed onboard as we are sailing. We have used our wooden scoop to bail water out every now and then since we started. That job kept us in constant motion and helped keep us warm. Our fingers were swollen and blistered from holding the sheet, and they hurt. Only by using all my willpower did I manage to set the sail on this morning — my muscles did not really want to cooperate. With two reefs in the sail, we were still doing good speed. The sun was not visible, so we had no idea where we were heading. Was continental Europe straight ahead hundred of miles away? We finally did get a glimpse of the sun and even managed to register its culmination at noon. We were well south of yesterday's latitude and were at best heading southwest.

Should I turn around and go north, or should I continue in the direction I was heading? Intuitively, I felt the Vikings would have pushed on, and we would do the same. The problem was that I did not know exactly how far to the west I was. But the good sailing at the beginning of the voyage must had brought us half-way to Shetland. We were still making headway to the west and were presumably well over on the British side of the sea and should see land soon. The second night went by rather slowly. We finally tied up like the night before and slept, but just for a short time. Then the cold came creeping and woke me up. Slowly, a gray morning light came back. Stig was awake in *Frøya*. He was also freezing and we decided to move on so that we could set our bodies in motion and generate some heat. Everything on board was wet on the outside. The wind was gentle for the full sail. With the wind blowing from the west, we fought for every inch, and continued sailing southwest. The day was dull with tons of clouds hanging overhead. Everyone missed the sun. It would have been so nice: warm weather and a calm sea with fishing, sun-bathing, rowing the rubber-dinghy, taking pictures, an above all, drying our clothes and preparing the first real meal since we started. Not this time. We were bestowed only cold, wet and rolling conditions. Observation of directions was impossible. The wind alone was completely in charge of our sail. That was the only element that can take us to shore and it was also the element that controlled our day on board the longboats.

We, the Navigators

My original plan was to start the voyage further north on the coast, but I finally decided to start right north of Bergen, which was much shorter. A steady wind from the east would have worked well with my plan. But with the prevailing wind blowing from the west, Stadt would have been a better choice. Shetland would have been easier to reach. At the same time, we also would have had a greater chance of bypassing the islands and continuing out into the main body of the Atlantic Ocean if we started too far up the Norwegian coast. I did not want that to happen. But the main argument that settled the case of where to start was the fact that some of my crew members had limited time and had to be back at the end of their vacation. So we picked Solsvik and hoped that the easterly winds would stay with us for at least a couple of days once it came around. Unfortunately that did not happen.

"Now we are heading for Scotland," I told my crew on the third day. Our sailed distance must be more than the distance between Bergen and Shetland by then. The previous night's observations indicated that I was still further south, which I fully expected. We also experienced difficulties with the *Frøya* keeping up. That added an extra dimension to my navigation. I know the speed of my boat, but could not tell exactly how much time we were loosing by stopping or slowing for our follower to catch up. Sometimes, that forced us to steer more to the south than we planned to.

I finally decided that we missed all the islands. Hopefully we are heading for the Scottish Mainland, I wrote in my logbook, adding that the most possible landing place would be Aberdeen or Peterhead in Scotland. If we overshot, we would have another one or two days of sailing and could make land at Lindisfarne. We were more than exhausted already and did not want to spend more time in this miserable environment if it could be avoided. All of us lacked sleep, a good meal, exercise, rest, dry clothes and a hot bath. We were physically and psychologically exhausted.

The next day we reached land. It was Scotland. The old and majestic remains of the castle at Cruden Bay greeted us. It looked like it was attacked and sacked by Vikings not long ago, but that was an illusion. The port of Peterhead became our landing in Scotland and a supply ship aided us in. Peterhead suited us fine, as we later planned to go to York. That upcoming voyage and the sail back to Norway is another story. Our relief was great when we finally got ashore. The crossing was so much tougher than we expected. But we made it in an old boat that did not produce one single leak during the demanding crossing.

I was very proud that we sailed without any modern navigational aid or equipment that could ease and guide our voyage. The record for this voyage shows that we must have sailed well over 300 nautical miles from where we started in Norway. The voyage took us 66 hours. That gave us an average speed of 4.5 knots, which was very good, bearing in mind that we stopped for a few hours twice. The wind was blowing almost all the time and we were making 10 knots or more sometimes (Engvig 2000, 2001).

We made landfall not far from the ruins of the castle used in the early Dracula-film shot at Cruden Bay close to Peterhead, Scotland. However, the relatively swift sail was the most demanding and exhausting any of us had ever experienced.

Of all the places we could have reached shore on this voyage, it was interesting to note that the place where we landed was the spot on the British mainland that is closest to Norway. From Cruden Bay in Scotland to Jæren in Norway, the distance is 250 nautical miles across. Jæren is 120 nautical miles south of Bergen, and we started north of the city. In 1914, the Norwegian aviator, Trygve Gran, became the first person in the world to cross a large stretch of water in an airplane. He flew his tiny Bleriot-monoplan named *Ca Flotte* from Cruden Bay across the North Sea and landed in Jæren. At that time, it was considered a major leap forward in air navigation over open water. We sailed the other way. Hopefully we also made a contribution to further the understanding of navigation on the open sea.

There were several lessons learned from this voyage.

First of all, the new radio positioning system we brought for collecting sailing data on board the *Frøya* started malfunctioning during the first evening out and broke down the next day. It was supposed to work even under tough conditions. Our longboat probably became too much of a challenge. I predicted my standing position every day and recorded the information. This shows that I had a pretty good knowledge of where on the open sea we were. I also learned that one of the crew members in the *Frøya* had kept a watch in the pocket without anyone knowing. Stig proudly told me that on the many incidents when I estimated the time of day by looking at the sun, I had never been more than half an hour off. Sometimes I could be within 5 to 10 minutes. He was truly amazed and found it hard to believe that this was possible. The day before we reached land, I entered the prediction that it would most likely be Peterhead, Scotland into the *Hitra* logbook. This documentation alone points to the fact that skilled Viking navigators did not have to rely on or use sophisticated instruments to find their way.

This Viking expedition told us that the sun alone was the Vikings greatest guidance on the ocean for their voyages. Stars are of no use at these high altitudes during summer. We could see one planet and vaguely a couple of Alfa stars, but the problem with those few faintly stars is that they do not tell you which constellation you are looking at. At best, we could use the bodies to ensure that we kept a steady course. Then again, we knew that the course was steady in regard to the wind as we sailed close-hauled almost all the way. Formation of clouds near the horizon also gave a sense of direction. Rightly trimmed, a longboat sailing close hauled will steer itself with no use of the rudder. The fine adjustments are done with the sheet of the sail. Wind and clouds change over time. This left us with the sun as the sole dependable indicator for navigational purposes on these high latitudes during summer. Knowing the speed of your boat and the days of sailing gave us the mileage, or how far out or to the west we had gone.

The "season of light" would be the natural time for the Viking fleet sailing these harsh waters. I think the sun is a wonderful aid and I believe the Vikings knew how to use the sun very well, and they used it as we did to sail the high seas. The relative darkness during the short night hours was my best indication of where we were heading. It told me if I had kept a straight course to the west or if we were being forced off towards the south. The night became distinctly darker only a few degrees to the south of the latitude we started, just before mid-summer. It was easy to tell. Only if the Vikings had fog for days on end did they get lost. So would we. Besides this, the performance of the boat would tell where you would be and the expected time of arrival. That was crucial knowledge to the Vikings, as it was to us. Accordingly, the Vikings measured distance at sea in days of sailing.

However, one crucial fact remains, and it is so obvious that it is often not taken into consideration when scholars discuss Viking sailing and navigation. The wind and the weather are always the determining factors. No matter how good the navigation, the boat or the crew might be, when it comes down to practical sailing, it is the wind that controls where the ship is heading. That fact alone will determine if you will hit your target. With one square sail and a low sail/rigging, like what we had when we reefed-down and were doing rough sailing, it is hard to tack without drifting off. That means you only have a window of about 180 degree to set your course. Under optimal conditions with a full sail, our 19th Century-rigged boat can do 60 degrees to the wind. This angle gets wider as conditions gets worse.

Showing off square sailing, coming in from the sea on the West Coast of Scotland. It is no exaggeration that old *Hitra* does better than 15 knots. Notice that a considerable portion of her keel is out of the water and that the long top pennant is flat against the sail in spite of the strong wind. It sounded like we were riding on a pebble road, probably due to air bubbles underneath the flat middle portion of the hull.

The Gotland picture stones of Viking ships depict ships with sails much like ours when we have reefed-down. If those boats had sails like that, they would definitely make their best sailing performance with the wind abaft. If the wind blows right in your nose, there are only two good options. Pull out all your oars and start rowing, or turn around, go back and wait for the wind to come around again.

The Vikings stayed ashore and waited for the wind to be right before they set sail. If we had kept the wind we had when starting, blowing steadily out of the east, we would have made it to Shetland. When it swung around, and started blowing from the west, we were fighting a losing battle. That was why the *Hitra* is equipped with ten long oars. We did not bring rowers, and my final options became Scotland, if we did well, or a return to Norway. Trying to tack back and forth is not a tempting alternative. In such a case, we could be staying at sea for a very long time.

Some books on Vikings refer to a long list of things that helped our ancestors in finding their way — from drifting kelp, seabirds, sunstone to a wooden compass-disc. We can tell from experience that to sail for days under the conditions we had is entirely different from theorizing about viking navigation behind an office desk. None of the signs mentioned would have helped us in any way. We saw some oil installations and also ship traffic close to the coast but they did not yield any new information. We saw some seabirds, but they did not tell us where to go.

Our expedition to Scotland in an old longboat gave us a different perspective and a new understanding of the Viking world and the hardship they must have often experienced. I can only admire their seamanship, navigational skill and bravery. When you have learned your way by heart, you do not need a map. The northern ocean was the Vikings' home field and they knew every corner of it. Only the extreme voyages are recorded in the Saga. Even the first record of Vikings seeing the North-American continent merely reports that the ship was blown off. The leader knew he was too far to the west. He did not even bother to go ashore as he was heading for Greenland. So he turned around and sailed to Greenland. This sounds strange to many contemporary readers for several obvious reasons. But it sounds so plain and so logical to me trying to look at it the Viking way.

My final arguments for the simple way of Viking navigation in the North Atlantic and the North Sea is that the many excavations from this period has produced no item that could be labeled for sure to be instruments for navigation at sea. Countless numbers of items of all kinds that are associated with every aspect of life have been unearthed from this period. If they had used different and sophisticated tools for navigational aid, those artifacts must have been priced items and therefore should have been found like navigational instruments from later periods.

It is very hard to use any precision device in a boat that sits low on the water, which is constantly moving and rolling, and often drenched with salt water from the sea. Our two replica gilded vanes were useless as weather vanes for wind determination and only became nuisance. They were unlikely to have been carried on Viking ships at sea as my articles on the gilded vanes conclude (Engvig 2003). I strongly believe the Vikings sailed the seas on basic training, knowledge and experience in much the same way as we did. I believe they were given basic training since childhood, not unlike the one I had. The Viking navigator's guidance was the sun and his ever-changing "environment at sea," his knowledge before the voyage, the speed of his boat and days of sailing. The gods of wind and weather would provide divine help. They would see that the Vikings arrived safe and sound or were forced to take other options. Faith and a strong will was what guided their bold crossings of a hostile northern ocean.

Better or Worse?

Even if I have a lot of experience sailing open boats with square sail, I would never have guessed how exhausting it would be to undertake this voyage. After several long trips along the coasts of Northern Europe, I thought I understood Viking sailing and navigation. We do not insist that all the Vikings experienced what we did when we sailed straight out into the ocean. We selected rather small boats and a small crew with one crewmember disabled much of the way. That meant more hardship for the rest. But Vikings might also have had disabled or sick crew members, endured cold and wetness, and encountered pouring rain, high seas and heavy squalls for days on end. Even such difficulties as leaks, a damaged sail, a broken mast, fog and lack of wind could present extra challenges unexpectedly. From our experience on this voyage, we learned other lessons that few sources discuss because they have never lived through it. I will not describe our experience as living hell, but rather, a kind of self-imposed torturous challenge.

When we ended the voyage, we were completely worn out. The only things we could think of was a big hot meal, a good warm shower, a nice shave, new clean, dry clothes, and a long, good rest in a soft bed or relaxing in the sun, away from the rain in a wet, rocking boat. If Viking sailors who crossed the North Sea for 4 to 5 more days as we did had experienced what we encountered, they would have been in the same state of weariness as we were. We were extremely frustrated and tired. If Vikings had sailed on from higher up on the Norwegian coast and finally landed at Lindisfarne, that would have meant a voyage that was several more days at sea. The sight of a group of wild-looking and dirty men with long beards and bloodshot eyes would have been enough to scare the monks at Holy Island into bolting the doors of their monestry shut!

If someone had shut the door in my face when we landed in Scotland, I probably would have broken down the door to gain access to food, drink, heat and rest. Besides, it was the common Viking law to attend to people that had been walking the mountains or sailed the sea. The payback would come later. The Vikings' "Havamal" states that you shall provide food and shelter to travelers in need. Refusing to provide such service was considered very impolite and socially unacceptable.

The Saxon annals that recorded the attack on Lindisfarne only described how badly the intruders had behaved. If one only listens to one side of the story, the verdict might not only be biased, it could be entirely wrong. I believe that the attack on Lindisfarne could have developed from an unhappy episode, where misunderstanding and bad judgment grew into hostility among the Easterners and made them seek revenge. Could it be that the result proved to be something quite different from what was the original intention? The incident at Lindisfarne made the Viking's understand that this was a new way to swift and tremendous wealth. This part of history is not very well investigated.

That said, I am not trying to portray all Vikings in good light. Historical records clearly showed that they were bad too. Even the word Viking is believed to have developed from a word that means provocator. Groups or tribes might have started out by sacking and plundering at home and abroad, like the modern day gangs and militant terror groups.

In the Viking Age, as it is the case today, the exploits of the "bad boys," which may be a minority, commonly hit the headline news. The majority of the population could be living in lawful and mundane existence and never made the spotlight. The Saga tells of a limited band whose deeds were worth mentioning. Bad people always existed in all societies during all historical epochs. Greed, jealousy, religious differences and ethnic hatred drive people towards conflicts.

We, the navigators of the *Hitra* and *Frøya*, ask for deeper analysis of the Viking movement, taking Norse mythology, Viking law and customs, runes and religious practice into account. Also, the story of "The Nice Viking" that settled abroad should be emphasized, like it has been done at Jorvik Viking Centre in York.

It was a special moment sailing two Norwegian longboats up on the beach at Lindisfarne — the site of the first recorded Viking attack on the British Isles. This famous stone in the community's small museum depicts a Viking raid on England.

IRON RIVETS

Our most important objective was to test that the old and very flexible longboat, built with only inches of freeboard when ready for sea, would stay together and afloat. We had spent 50 days at sea in the *Hitra* before we took on the challenge to sail straight out into the unknown sea. We did 28 more days at sea after this crossing. The *Hitra* covered 3,700 nautical miles of water during all our expedition voyages. This is considerably longer than sailing the Viking route to Vinland. I had few doubts about the quality of my old vessel and expected it would keep together even during the very rough weather we encountered.

The feeling that the fastenings and the iron rivets would keep up and that the boat would be good enough was crucial for a success and yet so obvious. But we knew that on the high seas, anything could happen. The 5/8 inch thick and up to 12-inch wide boards in the lower part of the boat made for a very delicate construction. A single blow from an oar against the bottom could split a board and create a leak beyond our control. We could hit submerged obstacles or even encounter an aggressive whale. A kick by its tailfin would easily break up the boat. A freak wave could get us. A huge cargo ship could run us down easily. Or we could sail into a storm, the type that has drowned Vikings as well as hundreds of Norwegian seafarers and fishermen in open boats through the ages.

But we were well-equipped for almost every possible situation with emergency-beacon and other types of signaling devices. We had survival suits and two certified inflatable rubber dinghies, one in each boat. Our follower was a new boat that could easily carry all the crewmembers if necessary. Most of the equipment on board was made of wood and would all float. Usually, if such a boat rolls over, it throws the ballast and leaves the crew with a swamped boat with the keel up. If the mast is cut loose, it is possible to straighten up, and bail out an open boat with good luck and hard work. But that would be extremely demanding.

More than ever before, we can confirm why this ancient, yet so delicate, boat construction was never altered. Any alteration would make an inferior boat. Many tried and ended up with a less seaworthy vessel. The open longboat must have impressed generations before us, and it continues to do so. If the people on board knew their part of the job, the boat will do what it is supposed to do, or what it was constructed for. This Viking-type longboat was created with great respect for nature and to cooperate with the basic forces at sea. If these simple facts are kept in mind, the boat will usually do great and serve its occupants well.

Iron rivets made of good wrought iron will often be the last part to give in. A clinker-built vessel of this type will have the bands put in afterwards. That has always been the rule in Norwegian *klinka-baat* construction. When such a boat is wrecked or crushed in one way or the other, it splits up along the boards between strikes of iron rivets as the *Rissa* did. I have seen this many times. I have tried to break up an old boat for firewood by using massive force, but ended up only creating a mess. The custom is either to burn the old boat on a big bonfire for a midsummer's dance party at the shore, or to retrieve the boards and band to be used as firewood. With all the tar an old boat has received over the years it is very good wood for making a fire for cooking or heating. But it needs to be made into manageable pieces. Only after flipping off the roe and poring out the rivets to disassemble the boat in the same way it was constructed did I have any luck in making good firewood out of a boat. I am convinced that the Vikings did just that.

They retrieved whatever was left of the iron by dismantling the boat in the opposite way of building. They could thereby collect the iron for reuse and at the same time collect good manageable firewood for cooking their meals. Old boat rivets or fragments found in boatsheds or at archaeological digs seem to indicate that dismantling of old boats was done the way I suggested. It is particularly interesting to note that the excavation of the only Viking site in America, at L'Anse aux Meadows on Newfoundland, unearthed a Viking bloomery as well as a smithy where the settlers had produced good quality iron form bog iron ore with the help of firewood collected in the area. Not only that, it was further established that the Norse had been recycling old iron rivets at the site and also had worked trimming wood. A great number of fragments of rivets from boats as well as wood shavings were unearthed. To me, that is a proof that the Norse had a shipyard in America a thousand years ago (Engvig, 2001).

If the boat being taken apart is getting old, the rivets might have had a good opportunity to corrode. Sometimes, there would be little of the iron core left inside the rust. The rust will spread by about ten times its original amount into the nearby wood. The wet board would usually yield to the rust rather than crack. Not only are rusted rivets very hard to get out, those rivets will leave a big hole in the board. To plug it might even crack the board. Unlike boats made on a frame, to repair a lap-streak boat by replacing boards is a delicate and time-consuming job. Making such repairs on a lap-streak boat will often produce poor results due to big holes in the old boards. New rivets have to be set aside from the area. Also, a new board may easily produce new cracks due to different seasoning of the new wood compared to the old boat. I know this from attempts by people in my community where I grew up, and even learned it the hard way by experiencing it myself. It is therefore smarter to repair an open boat by riveting on doublings on top of the leaking area, and start investing in a new boat as replacement. When a considerable amount of fragments of ships rivets and spikes are present at an archaeological site in a setting like at L'Anse aux Meadows, it is consistent with abandoning of old vessels and the construction of new boats; and not the repair of old boats where only a few rivets are needed.

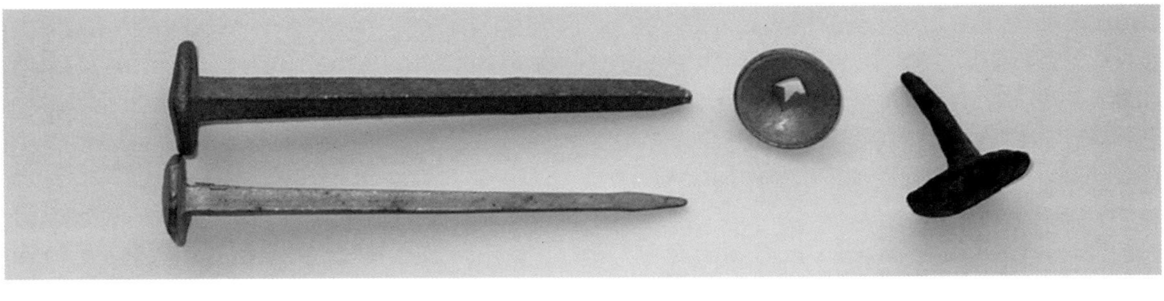

Modern and old saum.Upper a 3 inch rivet(shank, square, 0.5 cm at head), with a 5/8 inch rove. Below a 3 inch boat spike (nail) (square, 0.4 cm at head), all machine made. To the right fragment of a hand made rivet from the 20 feet long *faering Mr. Thomas*, built in 1875. Head width 2.0 x 1.5 cm, shank length 1.7 cm, diam. 0.5 cm.

The first full scale replica copy of the finest of all the Viking ship, the royal barge from Oseberg.

THE OSEBERG QUEEN

After my many Viking-way voyages, I decided to invest in a new boat as a substitute to old *Hitra* that had done such a wonderful job taking us along the Atlantic coast of Norway, around the North Sea, across to the British Isles and up the Irish Sea to the Shetland Islands before going back to Norway. But changing my focus from expeditions in small boats on the high sea, to the investment in a new boat became much more of a challenge then I ever experienced in ocean sailing with open boats (Engvig, 1987).

My dream, however, had always been to try to make an as close to exact copy as possible of the one Viking ship that has never been copied full scale before, namely the royal barge from the Oseberg find. This oldest of the relatively complete Viking ships would, as I saw it, be a fine follow-up of my research on wrought iron and how this barge could have been used. Many friends and family members wanted to support me on this, and as Sweden's TV 1 was planning a series of Viking programs based on books about the Vikings, they asked me to be a consultant in Viking-type square sailing.

Film manuscripts were being made and the Swedes were also starting to look for a full scale Viking ship to be used in some of the scenes. That made a good opportunity for seeking financing for this new enterprise. Banks made commitments, so we founded a company with the goal to build the first replica copy of what is believed to be Queen Aasa's ship from Oseberg. It was placed in the mound about 830 AD and excavated in 1904. The barge included a priceless collection of artifacts and artwork for every aspects of life with fertility as the primary message. It is perhaps the best known example of a pagan burial from the real Viking Age, giving us insight into the old religion and its splendid fertility cult offer (Ingstad, 1994).

The plan was to use this ship for the benefit of Norway and try to learn more about the Vikings and how their finest ship could have been operated under different conditions. A boat builder who was also a farmer signed the contract to build us the ship during the winter with little demanding work on the farm. Together with his sons, he built the full-scale copy of the Oseberg ship which I named *Dronningen* (The Queen) using oak as in the original and riveting the boards together with wrought iron rivets the farmer made himself in the farm's smithy. He made the rivets and cut more roes as work on the ship progressed.

During the barge's construction: the bottom was riveted together and the band just fitted on top of the fastening clamps. A four year old boy, Tormod, gives us a perspective on the size of the vessel.

The boards were cut out of thick oak planks, leaving notches or clamps on the inside for later fastening of the bands to the boards with baleen through holes in these inside clamps.

To acquire the baleen was the hardest part. Baleen whales were not hunted and the last ones left in Norway were from a breed of whales with baleens that are too short to be used in this construction. Fortunately, we managed to find the right type of long baleens in Iceland and purchased them at a very favorable price when they were told they would be used for the first exact replica of the Oseberg ship. When the bottom of the ship had been riveted together, up to the 10th board the assembling of the bands or the framing structure started. At this point, the ship looked like a flat barge, that was 70 feet long. A Norwegian Maritime Directorate's Officer visited the site of construction and supervised the building (K. Lein, personal communication, 1987). Lots of attention was given to the construction.

A local woodcarver was carving all the steam's ornamentation and scrollwork fore and aft. The boat's only "foreign item" was its name board. The serpent's head was even more accurate than the replica one sitting on the original ship at Bygdøy. New research on the remains of the original enabled us to make a better replica. It was a true copy of the Dragon-style decorative woodwork on the original. *Dronningen was the finest vessel made of wood in Norway in the 20th Century.* We also placed an order for a sail and got a rigger to prepare for the rigging. Hand-made lines were imported from Denmark. I followed the construction with several visits to the builder and spent much time working on the stability, trimming, mast, sail and rigging of the vessel. A friend who got the job of constructing some 30 copies of the ship's simple oak chests from the Oseberg found that they could be used as personal boxes and stations for the rowers when working the 30 oars.

The hull of the first full-scale copy ever made of the Oseberg Viking ship sitting in the boat shed.

Dronningen was the finest vessel made of wood in Norway in the 20th Century.

From my long experience with all types of smaller square sailing open boats, I knew that the most important part of testing it out would be the ballasting, mast, sail and trimming of the ship. Sailing an open boat with a few feet of freeboard is like sitting on a lever. When the wind kicks in, you have to be sure how much she will take before unwanted things start to happen. You have to know where your emergency options are and to apply them before things get bad. This is crucial for the ship to perform well, or sail close to optimal speed, and at the same time make a safe passage. I knew that the *Dronningen* would need to be well-ballasted and trimmed right to function well, so that it would not capsize.

There was a big celebration in spring when the first copy of the Oseberg ship was rolled out of the boatshed and moved down to the shore for its naming ceremony and launch. Cultural authorities from Oslo came and Norwegian TV conducted an interview where I outlined the goals with regards to research and the future use of the ship in education. The old boat builder had made a very fine replica. He stated that this would be his last boat. But unfortunately, he got personally attached to the boat and did not want to part with it —he wanted to keep it with his friends in his part of the country. Instead of finishing the work according to the contract, he sold our ship to the local people. When we learned that she had been removed by others, we tried to pursue the case but failed. To make it into a court case would cost a lot of money and could take years. We had no such means and lost our right and a considerable amount of invested money and labor with the loss of the ship.

To acquire a ship by taking it unjustly constitutes bad luck. On her maiden voyage, the *Dronningen* was loaded with only half the ballast I had calculated she would need. Sandbags were used instead of rocks, which are heavier and should have been her proper ballast. Furthermore, far too little of it was kept in the back of the ship. She was down 6 inches by the stern while I had calculated that she needed to draw at least one foot or more aft than at her bow. It is also a golden rule to start with more ballast then needed. It is easy to get rid of stones when you see you do not need them. Unfortunately, all of this was no longer in my control. As she was going under sail, a strong wind filled her sail and the person in command did not apply appropriate measures to let up on the pressure. She was forced over, pushing her forward side under. The inside quickly filled with water, and in a short time she was gone, just like the Swedish king's ship, the *Vasa*, 360 years before. She sank in rather deep water. After some time, she was located and a salvaging mission started. Still, with her sand ballast inside, she was hoisted to the surface and her side split open. Before she came ashore from her maiden voyage, the first new and beautiful Oseberg copy was a complete wreck.

Repairs and modifications then started. The ship got new strikes of boards on top of the gunwales so it came to resemble the Gogstad ship in an odd way. They made a hole in her bottom and installed an engine for motor propulsion, and modified the original fastenings. Sad to say, the educational and research goals behind the building of this ship along with its documentary film use were completely lost. So was the ship's authentic presentation and display value. The ship became a show-off piece for the uneducated with its nicely carved decorations. To make it even more bizarre, it got the new name *Oseberg*.

After she was rebuilt, this very odd-looking *Oseberg* ship was taken to Spain to be displayed at an exposition. On the coast of Spain, she foundered. She was crushed by a big wave from behind. The small crew of six managed to get into the rubber dinghy and was fortunately located by the Spanish Coast Guard and saved just before dark. During the maritime declaration following the loss of the ship, the Norwegian Maritime Directorates Chief Inspector in charge of the hearing conveyed strong criticism on the fact that no one aboard had certified maritime education required by most coastal states, or basic knowledge in navigation. He concluded that the ship's crew was not able to understand weather charts, nor did they seek professional advice or acquire local information on wind and weather before they left the ship's final port. With a storm approaching, the ship should never have left port.

This is the short version of a sad story of a very fine ship. At the ship's launch, when we were still in control, the media and historians declared the *Dronningen* to be the finest wooden vessel made in Norway during the 20th Century. Although this happened some years back, I stick to my old boats ever since. I never tried to make a new one again.

We brought copies of two gilded vanes from late Viking Age to test their usefulness. They were useless as weather vanes, but impressive as a symbol of status. They opened doors and made us special in more or less the same way as a mayor wearing his regalia today. We are convinced that the golden vanes were the symbol of kings back then. This is our copy of the Heggen Vane, from Norway, and it is covered with a coating of 24 karat gold. Ten years after the expedition it still looks the same (Engvig, 2003).

CONCLUSION

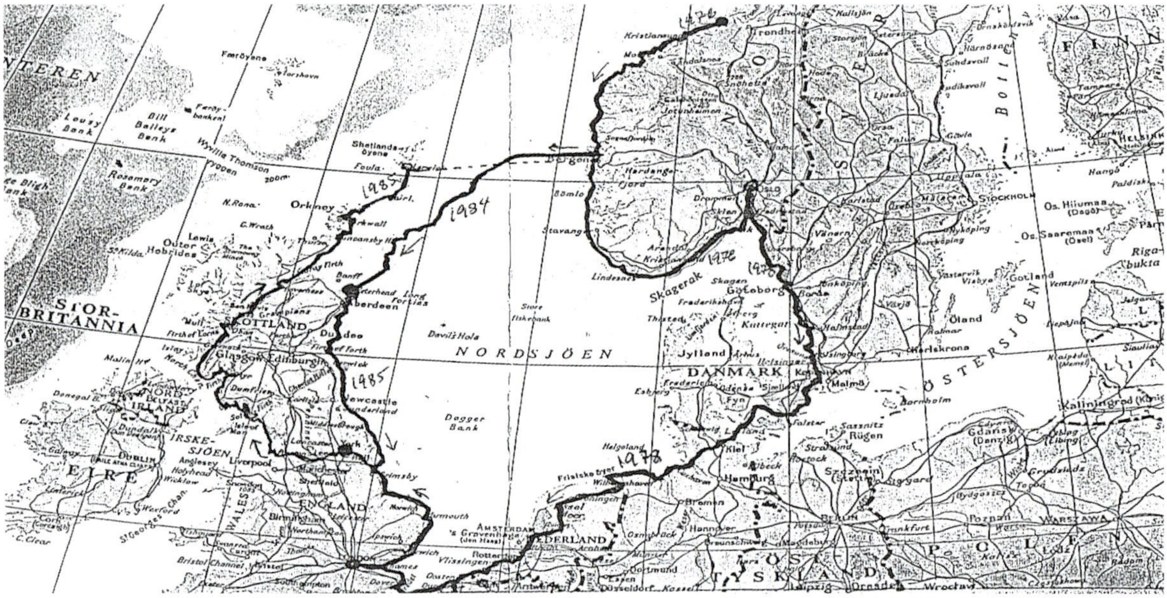

Together with family and friends, I made these voyages without map and compass to test Viking navigation. We brought traditional food and clothing, but also kept two inflatable rafts, an emergency transmitter and other safety and life saving equipment at hand in case of emergencies. The voyages are shown in this map. It illustrates all the four routes we took while testing out Viking navigation and the old open longboat's ability to withstand harsh punishment and do long and exhausting ocean and costal sailings (Engvig, 1978 to 1985). The recorded data is as follows:

*We spent 78 days at sea.

*We covered 3,700 nautical miles, which was longer than across the Atlantic Ocean.

*We visited eight countries, and called on 105 cities and township en route.

*It was all done in the original "Viking" longboat, the *Hitra*, built in 1863.

The 29 feet long boat was never restored. *Hitra* carries 10 oars and a square sail.

Frøya, our faithful follower

THE VIKING HERITAGE

When David Lewis wrote his books on Polynesian sailing and navigation, he based them on a living tradition that was on the brink of extinction. At the same time, half-way around the world, I started researching old Norse sailing and navigation. How could the Vikings find their way across huge bodies of open water and in a much harsher climate than in the South Sea? The Vikings sailed the northern waters more than a thousand years ago, basically with little or no extra help from man-made navigational aids. These two cultures of navigators would regularly set their course straight out into the ocean to small islands or masses of land far away. They used low riding open boats which were completely exposed, and the Norse sailed the harsh northern ocean. These voyages were undertaken at a time when most coastal people rarely sailed far away from the shore.

Two main conditions had to hold for any maritime undertaking of such a magnitude.

First, those mariners had to have a good and seaworthy platform for travel. The boats had to keep together for a considerable length of time and also be able to survive storms and harsh punishment from the waves and the strong wind. Above all, the boats had to work well in shifting winds and be able to make return voyages. This included periodically hauling the hull from the water and cleaning the underside to remove marine growth accumulated at sea. Marine growth slows a ships ability to make headway. Like the Vikings, Polynesians used to beach their boats. Even heavy Viking ships could easily get off again at the next high tide, so they would have a fresh boat for the return voyage.

They also had to have room for water, food and other supplies to survive for a long time at sea, along with confidence in their navigators, and seamen with the courage and skills needed to handle the ship under different wind and weather conditions during both day and night. They had to sail long stretches of open water where anything could happen en route. They had to play hide-and-seek with breaking waves, and had a reasonably good chance of survival as they undertook such voyages on a regular basis.

They came by sea. Leif Eiriksson was the first named man to bring European culture to the new world. He and his party stayed for the winter. His countrymen later developed a settlement in Newfoundland that lasted about 30 years. Copy of oil painting by Christian Krogh.

SHIPS, NAVIGATION AND MIGRATION FROM THE NORTH

More than a thousand years ago, the people living along the rugged coast and fjords of Norway with abundant supplies of fine wood for ship building and fine quality iron, had developed a unique boat. This was the lapstreak boat, sewn together with iron rivets rather than other material. It was light, swift and had a shallow draft, but was still a very seaworthy open boat. It had an optimal combination of sail and oars, making it superior to other vessels of its time in speed and seaworthiness under different weather conditions (Engvig, 2000).

Raised by the sea and with a culture that encouraged bravery, the Vikings feared none. They developed great skill in navigating the high seas without instruments. These Vikings were able to escape or challenge, overtake and capture. Even when cornered, they could get away by moving their lightly built boats over land. As a result, Vikings were brave and good warriors.

The Viking raiders, adventurers and immigrants were a confident bunch, who would venture far and wide in search of trading objects or new places to develop or settle. In doing so, they discovered land that was better than the challenging, barren and weather-beaten coasts where they came from. The Norse left marks in most of today's Europe, in Russia, Constantinople and even south to the Canary Islands. They ventured far into the North Atlantic Ocean to Svalbard, settled on Iceland and Greenland, and established colonies in the Americas.

Based on the fine seaworthy vessels they used, their social and adventurous spirit and the navigational skills they possessed, it is just natural to assume that they sailed beyond L'Anse aux Meadows to the shores and mighty rivers of the American continent itself, as well as the Caribbean Islands. These ventures would only be a mirror of what they did on the other side of the Atlantic Ocean even if the expeditions to the west from Iceland and Greenland were small scale compared to the expeditions eastward. There were no wealthy Arabs and favorable items to trade in the west, only primitive and hostile *skrælings*. So there were no strong reasons to venture in that region.

Recording of names on land or islands to the west, made before Columbus, only make sense if they are related to information from Norse navigators who described the land where they had visited. Columbus himself had knowledge of a land west of the ocean that obsessed him to making an expedition. His brave voyages would be compromised if it was known that he was merely following the footsteps of others before him. Columbus was a leader in need of success, but like leaders before and after him, he was inspired by earlier accomplishments. The same was the case for Leif Eiriksson who depended on Bjarni Herjulfsson's report of a land to the west. Even Bjarni might have had previous knowledge of the land he saw before turning back.

The coast of "Vinland de Goda" and islands far to the southwest were places where Norsemen would go. But since these places were already inhabited, the men from the North followed the usual pattern. They simply left the area, were killed, or made peace and blended into the existing population.

"The Nice Viking" – a skillful and hard worker who immigrated to England, Ireland, Normandy, Iceland, Greenland, and other places was probably much more common than mentioned in history books. *Skraelinger*, later named Native Americans, would probably have accepted some well-behaved young blonde Vikings, and invited them to settle with the tribe. They could even have gone to the extreme and destroyed their boats to make them stay. The Norsemen worked iron, and had artifacts and "tricks of the trade" that must have impressed the native inhabitants of the land. Compared to the existing population, the number of the visitors was few, and traces of their achievements are forgotten. Records are often lost, if not, they still need to be detected or understood.

It is challenging to prove that some Norsemen traveled to America. One day, laboratory research in micro-biological and gene technology, along with archaeological developments, will give us new evidence that may shed light on old maps and puzzling artifacts, and supply missing pieces to a lost history. This may eventually prove that some people from the North, in their swift and able boats, settled on the American continent hundreds of years before Columbus.

Even the Vikings who lived a thousand years ago found life to be easy on the new continent to the west. Leif Eiriksson was called "Leif the Lucky" because of the fine land he discovered. He gave America its first name, Vinland the Good.

Much later generations of immigrants were also drawn to this new continent. Many of them did well in America and we know their stories. Indeed, they proved the old Vikings right. Even if they did not sail in Viking ships, many emigrants from the North went there on a one-way ticket on ships chartered for the sole purpose of bringing new immigrants to America. Emigrants from other parts of Europe used regular packet or passenger ships.

At an early age, my father, the old master mariner, taught me how to navigate and tell time by observing the movement of the sun. Later, as a scholar, I learned from old fishermen that they hardly ever used maps or carried a watch or a compass, but sailed on the Atlantic Ocean sometimes without sight of land. When I decided to embark on my first voyage from Norway around the North Sea to York, I wanted to do it the old way as told in the previous chapter.

It was easy. Even a voyage straight out into the Atlantic to try to sail to the islands to the west was done in the same manner. We did it "The Viking Way," and experienced much of the same hardship that some Vikings must have experienced a thousand years ago.

We based our sailing and navigation on a tradition handed to me when I was young. We completed our voyages based on knowledge that, like for the boat we used, had been handed down over a period of more than 1,200 years. In many ways, my experiences equal those recorded by David Lewis (Lewis, 1972). But our environment was very different from the South Sea, and their mariners covered greater distances than the Vikings did. But like them, we found land based on the information we had before we set sail. In this respect, we accomplished the same task as Leif Eiriksson and many other unknown Vikings before and after.

250 nautical miles out into the ocean, the wind had changed west and picked up. We experienced some demanding sailing and some of our crewmembers started wondering if we ever would find land. What if we shot past the islands and sailed straight for America?

SHIPYARDS IN AMERICA

The Ingstad excavations at L'Anse aux Meadows on Newfoundland showed that the Norse settlers, more commonly named the Vikings, established a settlement at this site around the year 1000 AD. They constructed a bloomery and produced iron and had a smithy for further work on the product. Most importantly, they set up a yard for ship building and ship repairs.

The fact that the Vikings used iron rivets in their lap-streak ship construction enabled them to make light and swift vessels, and they traveled far. But a ship's life could be short. The vessels always needed maintenance, and the fleet needed constant renewing. That is why Vinland, more than 500 years later renamed America, was such an important destination. The people in Greenland had two options: go the long way to Norway and buy vessels or import prefabricated ships at considerable cost, or go to America and make them.
The latter must have been the cheapest, and for many settlers, the only affordable solution. The Vikings were probably the first Iron Age men to establish production sites of an industrial nature in this primitive world out west.

The native inhabitants outnumbered the Vikings, but they had no use for what the Vikings needed the most. Mossy moorland with bog iron ore and wooden areas including old pine trees were what these Europeans were looking for. It would not be found in populated areas, but rather in the backlands where the newly arrived artisans could go about their work without being interrupted. After the first discoveries, the Vikings came back to set up iron production and a shipyard. They repaired and renewed their sailing fleet in America, where ship building materials were abundant. That is probably why the land was named Vinland the Good. The Newfoundland to the west had the key raw materials to sustain the community in Greenland (Engvig, 2001).

The following will focus on the Greenlanders' need for ships and supplies and why the skill of making iron, logging and making boats were closely-linked to one another. Their skills and ventures were crucial in maintaining a good standard of life in their cold outpost named Greenland. Their search for new opportunities led them to aim their exploration westwards. These were the only unknown areas to the mariners that emigrated from Norway.

My proposal is that the Vikings' voyages west were not to "raid, pillage and plunder" or to trade like they did in the rest of the known world. What could these Iron Age men be trading with Stone Age people that lived on a similar land and had more or less the same living environment as they had? The native *skraelings* basically lived in the same way as the Vikings' ancestors did in Norway hundreds of years earlier. The Greenlanders were exploring the west for other reasons, first and foremost to find resources needed to build and maintain their fleet of different size longboats and their colony in Greenland.

Greenland was a good place to live for the immigrants from Norway and Iceland. They had good houses, domestic animals, fish and game, and plenty of marine mammals to hunt. They had an impressive natural habitat with pastoral scenery and all the things they needed, and we believe they had no hostile neighbors. They hunted and traded local goods and priced items with ships that came from Europe. But they lacked a few very important items that could not be shipped overseas without great effort and cost. Ship building timber and wood for construction were in high demand and in short supply. In order to produce the iron they constantly needed, they had to seek out places where fuel was plentiful. The women would not likely accept that the men use the winter fuel, gathered for heating and cooking, to make iron. To make iron, the Vikings needed lots of particularly good firewood that could produce the right heat in the furnace.

These materials were all found in Vinland, a considerable way to sail, but far nearer than to their former homeland. The way back to Norway was impeded by the huge and often hostile North Atlantic Ocean. The way to Vinland was coastwise sailing a great part of the way, which could have been done even in smaller boats. According to the Sagas, the common way to Vinland, as recorded during the Medieval Age, was to go north along Greenland and cross over the Davis Strait to Baffinland and sail south from there. The Vikings who came to the East Coast of Canada a thousand years ago would likely construct iron production plants with furnaces where good fuel for this enterprise was abundant.

L'Anse aux Meadows gives us the clue. This is similar to what happened in the Pittsburgh area some 800–900 years later, where coal supplies for a large-scale iron and steel production existed. The Great Lakes became a major ship building district and Detroit the city where early automobiles were made. The same pattern would also apply to Great Britain and the development of an iron industry based on coal. This will be highlighted and discussed in the next chapters.

For people living on a remote island like Greenland, it was crucial to have good ships and boats. They needed numerous classic Viking *faerings*. They were always used for trips along or across the fjord or to the islands for gathering driftwood, fishing and hunting, egg and down collecting, socializing or other tasks. The *faering* was the Volkswagen of the Viking Age. In Norway, it was used as such continuously until well into the last century. For coastal visits, expeditions and longer voyages along the coast of Greenland, bigger boats would be employed. They were also needed for collective transport or heavier work, such as ferry herds of domestic animals over to a green and grassy island in the fjord where they could get fat over the summer. For ocean voyages, the largest ship in the community was needed. For trips to Vinland, they could have used bigger ships but also taken medium-sized boats along to repair old ones and built new ones at the destination.

Logging, lumber splitting, woodworking, iron making, blacksmithing and tool making, as well as ship repair and ship building, were key activities young boys learned while growing up. Several persons in the community had to be proficient in these skills. Logging and ship building in the Viking manner has been handed down to new generations up until the 1900s in Norway without documented instructions. The basic technique was this: a "slow working" older man and a young student were tasked to build a *faering* together. They always start building from opposite sides so that they would not get into each other's way. The youngster had to copy the old man exactly for the port side to mirror the starboard side. In some districts with good ship building material, most of the farms would be making faerings and bigger boats on request, even producing the iron rivets and roes in the farms smithy.

Finds from the rich Viking dig at Coppergate in York. Three whetstones in the small box came from Eidsborg in Norway. They were crucial for sharpening iron tools.Whetstones from the same quarry are still in use today. We had one for sharpening our tools on board the *Hitra*. Finds from Coppergate are now on display at the Yorvik Viking Centre and Museum.

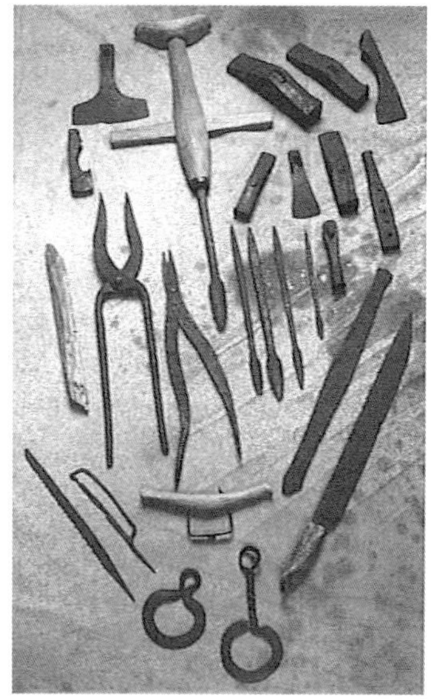

A smith's tools from a Viking grave. Several such finds has been unearthed in Scandinavia. Much the same tools are used in traditional smithies in Norway even today. Photo MNA, Sweden.

The blacksmith was the wizard of the ancient Norwegian world. His work was an art infused with myths and superstition. In fairytales he was the only human that could play with the devil. He was obviously an important person. Knife-making the Viking way is still a distinguished art in all of Scandinavia. For the iron production processes, the same intricate and exact knowledge had to be acquired by each new generation. Not until 1782 was the ancient bloomery iron making processes recorded in writing in Norway by Ole Evenstad (Espelund, 1999). The building of smaller traditional long boats of the type presented in this book was not recorded properly until our own time (Engvig, 1969–75).

On the northern tip of Newfoundland, the Vikings established a permanent community where between 50 and 100 persons are believed to have stayed over a period of 30 years. Many of them would likely have been engaged in ship building. Lots of shavings and old broken rivets found on the excavation site indicates that they worked on boats. In modern day terms, they would have been shipyard workers.

The area surrounding L'Ance aux Meadows would have had good ship building material. According to scholars, this was also a Viking base camp. Norsemen departed from this spot to investigate the area to the south as has been shown from the finds at the site. Others might have been passing through on their way to and from other locations along the nearby coast. They would use boats as they were the best and safest means of transport along the seaboard and rivers of America. All these activities creates a high demand for new tonnage.

L'Anse aux Meadows might well have been a Viking encampment, but the area would soon have been drained for its bog iron ore. Iron searching teams would set out to find new locations to gather and produce more iron as an alternative to costly import from Norway or Iceland. L'Anse aux Meadows was abandoned for unknown reasons. Since their homeland was Greenland and the people at home were waiting for "the new boat" or more supplies of iron or other goods from Vinland, they would probably set up temporary camps at various production places. Several bloomery production sites in Norway from the pre-Viking, Viking and Medieval ages have few remains of a major settlement. It is the factory site and the different furnaces that make up the distinctive remains, not to mention the slag heaps. After the bog iron ore was utilized or the fuel supply was exhausted, the Vikings would move on and leave the slag heaps and the round indents from the furnaces in the ground as evidence of their activities. Yards or sites for the building of boats and ocean going vessels leave fewer distinct visual remains of such an activity.

The Vikings are believed to have been too few in number to colonize Vinland. Furthermore, the land was already inhabited. Any settlement had to be defended against hostile takeover. Also, a preferred area for settlement would probably be where other people would like to live, and where people are living today. If they ever tried to settle in a popular area, which might have happened, any distinct Viking remains would likely have been destroyed long time ago.

About a thousand years ago, this site at L'Ance aux Meadows was inhabited by the Norse. Excavated by the Ingstads', it contains the earliest record of iron being produced in America.

Woodworking and ship building in America a thousand years ago with a smithy and a nearby bloomery by the brook. The camp is made up of both temporary and permanent dwellings and a dense forest with good wood can be seen in the background. Ships are coming and going. Depicted by fourteen-year-old Håkon after a visit to L'Anse aux Meadows, so far the only documented Norse settlement in North America.

When looking for new Viking sites in America, we should not primarily try to find other settlements like L'Anse aux Meadows, but look for remote areas that would have had good pine forests with moorland where bog iron could have been extracted from the mossy ground. That would be the natural production site for a Viking Age bloomery. Ships could be built at any suitable location near the water with good ship building materials close by.

The Iron Age Norsemen would come, stay for a while and leave. They were always on the move. That is part of the Norwegian spirit even today. It is a culture characterized by restless and adventurous people who like to explore new places. Exploration of the new world as we know today started with Leif Eiriksson a thousand years ago and it ended when the whole planet was mapped. The last great discoverer was Roald Amundsen. He mapped large portions of the only white spots left in our world, namely the Polar Regions. Between these two explorers who lived 900 years apart, almost all the great exploratory discoveries on this planet took place.

I believe there were more iron manufacturing sites and Viking shipyards on the American continent than the single one that has been partly excavated. Hopefully, there are located in more remote areas that have not been compromised by development. The preferred bloomery fuel was a dry pine tree. The furnaces were set up along a bank by the water where fuel was plentiful and iron ore could be found and where they could haul ships ashore close by. A camp at such a site could be quite primitive and leave few traces many hundred years later. But the round impression from the furnace and the slag heaps would be there. They would be genuine proof of the Vikings' activities. It should be noted that the further north they could work, the closer to home they would have been. Pieces of forgeable iron, remains in the slag, odd broken tools or discarded old rivets could be signature items to look for. Such iron fragments could still be present, as wrought iron remains is a relatively lasting product.

The previous arguments would also explain why the Vikings named the land: "Vinland the Good." It gave the Iron Age men from Europe the thing they needed the most for everyday life in Greenland — iron. Sharp iron tools with steel edges were needed in everyday life. Iron tools were used to cut the grass and the trees, split logs and built ships. They were used for protection, for hunting, on the farm and in the kitchen. The Norse used their iron tools to make wonderful artwork, superior ships and sturdy houses. They were the friendly Vikings who sailed to Vinland.

Further research and archaeological investigations are needed to find new pieces that would fit in the puzzle and throw more light on the first modern men on the American continent. The Easterners came by sea in open boats to secure and explore possibilities of a better life. For that purpose, the continent proved to be good. That is why they established the first modern shipyard for ocean-going tonnage on the American continent a thousand years ago. It is tempting to add that the big wave of immigration from Europe that started after the rediscovery of the continent 500 years later were boosted on much the same motives.

The Norse settlement in Greenland was founded before the year 1000 AD. It existed for almost 500 years. Living on an island, boats would be in high demand. Average Norwegian longboats will usually last 30–40 years if well-kept and maintained, sometimes even longer. Many boats had a shorter lifespan. If left exposed to frost, rain and the sun, a new boat could deteriorate beyond repair in less than 10 years.

We could simplify all this and state bluntly that each new generation needed new boats. As far as we know, there was no significant export of boats from Norway to Greenland during the Medieval Age. To supply the community with this crucial item it would have been natural to use the resources in America for this production. Boat-making, as well as iron production would, in my opinion, be a main reason for the Greenlanders to keep in contact with the New-found-land to the west.

The sleek lines of *Mr. Thomas*, a traditional 19' Norwegian *faering* or "Folk-boat." Built in the 1870s it belongs to the last generation of small craft built when people were still dependent on them for transport and work. The curved lines of this construction have become the symbol of the superior iron riveted lap-streak crafts that the Vikings used to roam the northern sea.

WESTWARD BOUND

The fact that the Norse went to America is a key to further research. In order to understand why they went there, we need to recognize their underlying motives. It is not this book's intention to speculate on what they did or did not do. I have instead tried to highlight some basic needs and demands the Greenlanders had. For the relatively isolated communities in Greenland to get important supplies, they would need to travel to Norway, the British Isles, or go west to America. To do the long haul east, across the Atlantic to Norway, would demand ocean-going vessels. That type of ship saw great development during the period. Instead of using open boats to sail on the high seas, sturdier decked vessels were developed and employed on ocean trade routes. They protected the cargo much better, were far more convenient and could carry a greater load than the open type vessel the Vikings had used. As people from Europe traded with Greenland, we must assume that sooner or later the community would lack the big ships required for that journey and rather become dependent of arriving vessels.

Voyages the other way was less demanding and could be done in smaller coastal ships like their traditional open longboats with square sails. But even this type of vessels needed to be renewed. It is hard to understand that Greenlanders would stop going to Vinland after less than half a century, when we know that the Norse people continued to live on the island for more than 400 years. There are several obvious reasons why the recording of further voyages to Vineland is missing. The most obvious explanation is that the local people wanted to keep this information to themselves rather than share it with everyone else. For the Saga writers, further stories of many eventless voyages to Vinland after Leif Eiriksson's landing were perhaps too dull to be told. That is, indeed, consistent with sailing on other trading routes.

It is not always wise to tell everything to the information collector or saga-writer. Many examples from small and isolated communities show that to be true. If someone had secret places or hidden assets, it is natural not share it with others. When the iron vessel *Hansteen* did its first depth soundings mapping off the Norwegian coast, the ship ran high aground with local people on board. That rock was a secret fishing ground. Even when mapping from small open boats, the local fisherman rowed the official surveyor in circles around a shallow area where he had his best fishing place. It was not for the stranger to know, and put that in a map for everyone to see. Only with the use of more sophisticated equipment was the truth revealed, and historians today can compare maps and see how the fisherman tricked the hydrographer into not registering the submerged rock. So, going by the same logic, why should Europe know what the Greenlanders had found out about the new continent?

Good iron was a key to the building of lap-streak longboats. Wood was the other. The early site on Newfoundland shows that the Norse had used that area, and tried to produce iron. Research so far has no answer as to why they might have struggled for good results and produced the superior quality that is a trademark of Viking and medieval iron from Norway (A. Espelund, personal communication, 2004). This again, shows that there was good reason for the Norse to hunt for new locations where they would try again for better results. They had ship so they had the means to move on. It was likely that the southern expeditions along the coast and rivers that we know of, were, among other things, on the lookout for new and better places to produce iron and build new tonnage. Expeditions would definitely have checked out the area to the north along the way they came. They had the knowledge and would set up a camp for making iron if they felt the results could be good, and if they could go about doing the job without to much interference from *skrælinger*. Looking for iron-making sites along the coast of Labrador would be an archaeological challenge (Engvig, 2001).

Iron makers and blacksmiths are highly skilled artisans. It was natural for a good blacksmith to keep several young boys hanging out at his premises to pick up the tricks of the trade. The same can be said about at the bloomer furnaces.

Youngsters would aid in getting wood, helping out and learning the trade until they knew it by heart. The Norse had no strict guild order like in feudal Europe during the Medieval Age. The system was more like the open American system that developed on the continent many hundred years later with artisan workers and youngsters learning by participating and aiding in the production (Gordon, 1996). It is tempting to draw parallels between the Viking iron makers on the American continent and the modern European-American that came and made iron many hundred years later. They used much of the same techniques of learning, which was rather different from the old central European arrangement.

The Gogstad Ship has become the prototype of today's replica Viking ship. Built around 900 AD of oak and iron, it was 23.3 meters long (76.4 feet) and 20.2 tons. It had 32 oars and must have been a speedy and seaworthy vessel as the many replicas has shown.

A Viking boat can be built in a surprisingly short time when all the material needed is at hand and good craftsmen are supervising. Most of the labor-intensive operations do not require outstanding craftsman skill — just average handiness, and familiarity with the different tools and the work operations was needed. Many adult males would have all these skills. In a society like the Scandinavian and the Norse colony in Greenland, young persons would be trained to acquire those general skills. They were a necessity in every-day life on the farm or in the household that often would be located some distance away from the nearest neighbor or settler. As far as we know today, Vikings did not built any major cities in Norway during the Viking Age but preferred the solidarity of the lone farm. However, when it came to the need of pulling together, they were masters. This is the typical trademark of a maritime culture. Teamwork is needed to launch and operate a Viking ship at sea, or to be successful in pillaging and plundering which has become the most significant trademark of the Vikings.

However, the Vikings did settle in numbers in old foreign cities like York adapting well to city life. But for the average Norwegian, like those who moved out to Iceland and Greenland, farm life on a self-sustained unit in the countryside was the life they were used to and brought up with. Villages and townships with the farmland all around the community were typical further south in Europe. The most typical issue in the Norse way of life was the dwellers' feelings of independency. The farmers were free and independent but clearly understood and developed the need for working together to the benefit of the whole municipality.

It is no overstatement to say that the Norse were good craftsmen and artisan workers, well-trained in different "state of the art occupations" needed in the society and especially in the art of working with wood. It has been stressed that the boats and boat building would be the number one goal to achieve, living in Greenland. Without boats, the community could not thrive. My knowledge of traditional boat-building in remote places, like Aafjord used to be, tells me that a team of men in the search for new possibilities in Vinland would include someone that knew how to produce iron, boat builders and other handymen, sailors and farmers. This expedition of maybe 15–30 men could easily build an ocean going boat of the Viking type in a month or so with the iron and wood at hand. If they selected a place where the nomadic natives were gone for half a year, or found a place with non-hostile tribes or no natives, the Norse could have had plenty of time to build new boats for the return voyage and split the crew. As stated earlier, smaller and intermediate-sized boats could be built more quickly.

A lap-streak Viking boat could be repaired like a carvel-built boat by replacing streaks of boards, but this was a rather complicated operation compared to replacing skin planks of a carvel-built boat constructed on a system of rather massive frames. If a Viking boat was beaten up, it would usually be at the shore or close to land. If it breaks apart on the high sea, it would not reach land to be repaired. Most often repairs on a Viking vessel was needed because the vessel was aging. In those instances, the owner would try to mend the boat temporary and in a simple manner to get a few more years out of it. He would make simple temporary repairs as could be seen on old Viking ships and a lot of aging traditional Norwegian boats. The "royal" barge from Oseberg has repairs, and is said to be a well-used and old vessel, probably not a good seaworthy ship anymore when it was placed in the mound.

Today's technology in dating objects of wood has established that the Oseberg ship was placed in the mound in the fall of 834 AD and that the ship itself was built about 820 AD. If this extravagant and costly barge of oak was an old ship at the age of 15 or so, one would think that an every day workboat would last even shorter. Even if this assumption is not quite right, it is likely that it must have been a great production of boats all along the Norwegian coast during the Viking Age. A thousand years later, this was business as usual. Fish was always a great article from the north that could be caught and traded. The sea was full of fish and boats were needed to harvest this delicacy. Farms had a smithy that produced iron. Every farm had tools for working the woods and further woodworking. They would make whatever needed from wood, even producing fine artwork and decorative artifacts for their house of worship, as well as other important buildings and ships.

Boats and seafaring, statesmanship and organization of new colonies, *all-ting* and free men; all are clear evidence of a developed society. At home the splendid Oseberg find shows us that early Viking Age had a highly-developed culture, including arts and crafts that came second to none in the 9th Century. Some scholars even assert that this heathen culture was clearly superior in many ways and achieved more than other European cultures. The barbarian Viking's key asset was the boat, from the greatest long-ship to the smallest rowboat. Every farm that needed boats and had access to material could make them. The building of boats must have been far more common than repairing or trying to mend old boats that had already started to rot or deteriorate.

To make a lapstreak open boat is simple, but to become a good and able boat builder takes time and skill. It is natural to assume that vessels like the Oseberg ship would require able boat builders and woodworkers. The Sagas even mentions this skill with a sample from the building of the *Ormen Lange* where the boat builder put his life at stake trying to make a more beautiful ship for the king. It is a fact that even in the 19th Century, several of the 40–50 feet boats, built for the winter cod fisheries, and of a similar design as the Viking long-ships, were poorly built. They could be built by farmers with less skill, while other well-made boats had a superior hull and would sail and row well. Some of the inferior boats would even get derogatory nicknames, but odd as it might seem, they often kept together and served their owners well. This is still a fact with regard to Viking ship replicas.

The Norse would continue travel to the west before or after the settlement at Epaves Bay had been abandoned. Going to Vinland would be the natural way of travel for adventure and supplies. Their light vessels were the perfect ships for discovery sail along the American coast and along inland rivers. It is obvious to me that people from the North could have made furnaces in Ohio (Mallery & Harrison, 1979).

These islanders living at "the end of the world," as seen from Europe, had good reasons for sailing west as this book shows. To them Greenland was the nave of their world. For historians of today, the area has become much more of a world meeting place than the old Europeans probably saw it. It became the first place the runic people from Europe, with their roots going back to the Indus Valley and beyond, carrying their contemporary European culture, met with people that came the other way having emigrated from Asia. These two very different cultures met when Vikings and *Skrælings* got together a millennium ago. Norse and Inuits continued to have contact on this Arctic island and in North America for hundreds of years before Europeans landed in the Caribbean.

Communication problems and different levels of development probably made it difficult for the Vikings to establish good and lasting relations with nomadic groups in the new land. The Norse, raised in a farming community, would naturally try to go about their business without interference. The natives must have seen the intruders and their mysterious use of the land as possessive and frightening, and this could perhaps led to clashes. Some Norse items have become possessions of Native Americans and ended up in their sites. A planer with missing iron tells us that the natives only had use of the iron.

While hoards of artifacts from the east have been found all over Scandinavia, little from the west has showed up, suggesting that trading was not the important factor for the Norse in America. The conclusion would be that farming livestock, making new houses or farms, and building boats were important issues in the lives of the Arctic Norse people, apart from maintenance work and harvesting from sea and land whenever the time was right.

During the Medieval Age, ship design and construction saw a great development in Europe. Ocean-going ships were constructed with decks and stern-rudders instead of the ancient steering board on *styrbord* side. They developed superstructure fore and aft, multiple masts and several sails. Those were the new and comfortable ships coming to Greenland. It made no sense to try to sail to Europe in open boats like the Vikings did, so the rather conservative Greenlanders became dependent on the ships from Europe. But for voyages to the west, the open boats could still be useful.

Like the river system in Eastern Europe, the waterways in America were well-suited for light open vessels with a shallow draft. The new continent had old-time highways in all the lakes and rivers that criss-cross the land much like Europe, only bigger, better and easier to navigate. This was the perfect highway system for boats like the Viking longboats. With a small fleet of such vessels, almost no part of the American continent would have been inaccessible to the Norse. Between one and four feet of water was all that was needed, and if the boat struck bottom, they would move to one side making a list that will lift the keel off the bottom. Then the boat can be pushed off by the oars. The longboat was a superior craft to any native vessel.

The Medieval Age also saw the development of new forms of sport and adventures. What if adventurous young men from Europe rewarded the Greenlanders handsomely to give them information or to guide them on a mission to the new world to the west? We know that the clergy and the king supported such expeditions. The great trend in those days was to embark on grand ventures, the most famous being the crusades to the Holy Land. People from all over Europe went on journeys to fight for Christianity.

It was likely that some would try to seek adventures in Greenland and the west where there was plenty of room to explore. That land was not densely populated and their inhabitants only had stone tools and simple boats compared to the Norse. Encounters with *skrælinger* could be more exciting and perhaps not so deadly as the Maurer's and other enemies to the crusaders. Some young Europeans had all the means to undertake such voyages. Scant sources seem to indicate that some Europeans did continue to go to North America before Columbus. If they did, they represent the story of lost America, a story still waiting to be unveiled. If people going there ever recorded the journey, it has been lost or is waiting to be rediscovered or interpreted. New research on missions supported by the church to christen the poor native people of the new world is a tempting approach. The unsolved question is still the same: what really happened to the Greenlanders, and other visitors or adventurers that came along the northern route to North America during the pre-Columbian era?

The biggest of the small boats of the Gogstad find. It is about the same type and size as our own *Frøya*.

THE LONGBOAT

The Viking Age saw a whole range of iron riveted open vessels for all purposes of transport along the fjords and islands throughout Scandinavia. These ranged from the small and basic *faering*, the local ferries, fishing boats and the intermediate-size vessels for longer voyages, to the merchant *knarr*, the royal barge or the real long ship, the Viking Age man-o-war that carried names and labeled *Drake* or *Langskip*. Viking ships came in all sizes, but not in all shapes.

Over the ages since then, the term longboat has often been identified with the larger inshore descendents of the true Viking ships. That is a lap-streak open boat, the all-purpose vessel or "lifeboat," developed from the boat the Norse brought along wherever they settled. The basic fact is that this type of a vessel is the ultimate construction that combines strength and lightness and the ability to be handled by sail, oars or by men and horses under all thinkable situations. This boat could serve as a good and dependable traveling platform on any stretch of water, long or short, and even across land. Besides size and usefulness, it was also the least expensive vessel and the easiest to build and maintain.

Longboats are identified by their length, which is the common feature in all Viking boats where the length-to-beam ratio is distinctly different from the so-called round boats, found in other areas and developed for other needs and concepts of travel on water. A longboat often has a length to beam ratio of four-, five- or six-to-one, sometimes even more than that. It is definitely a long and rather narrow vessel. It should be added that even if a majority of all types of small crafts in the old days were clinker-built first and second cousins to the lap-streak longboat, a percentage of what could be called a longboat could be carvel-built and closer related to traditional fishing vessels from southern Europe. They would have more frames and be much more heavily-built and not so easy to handle.

We know that the people in Shetland had boats built in Norway. But the islands to the west also had famous local *batasmid* or boat builders (Petersen, 1966). It is interesting to note that the word means boat-smith in English, which clearly shows that the trade of boat-building had to do with shaping in iron and wood. Some of the islanders went to Norway and learned the trade there. When they came back, they needed apprentices for help and other locals were educated in this way. Descendents wrote books about their boat-building knowledge and their work building boats in the Faeroes. They made their own local adaptations or modifications to the boat design from western Norway.

In Ireland, the famous sailing boat back in the old days was the Drontheim boat, a lap-streak longboat imported from the Trondheim area in Norway that gave the boat its distinctive name. Its appearance is close to a traditional lifeboat that is carried on all sailing ships and engine propelled vessels alike. Like other smaller crafts of the British Isles and Ireland, it is a descendent of the Viking boat (MacPolin, 1999). The Irish of today like to point out this particular relationship, and that the clinker-built boats came to England, Scotland and Ireland with the

Vikings in the 8th Century. Their boat-building skills, as well as the particular way of construction used in the Viking colonies, were passed on to the natives. If we look further to the west, we find that many of the North American skiffs or dories are cousins to the Norse boat. The modified version double-ender used on the Grand Banks could even be stacked like chairs in the gymnasium. A thousand years ago, the Mediterranean people were impressed with the Viking long ships that could be rowed both ways. Their own boats, which were fitted with a rams at the front, were much heavier and had to be turned around, which was a cumbersome operation.

Georg Harboe and Frank G. Samuelsen were two brave but poor and humble oyster-catching immigrants who arrived in the U.S. from Norway. They grew homesick but could not afford to return. In 1896, they decided to compete for the grand price of 10,000 dollars put up by New York's Police Gazette for the first to row clear across the Atlantic. They were used to sitting at the oars in their oyster smack for long days on end and figured they could continue doing just that for the long Atlantic row. In 1896 they rowed from New York in an 18 feet open lap-streak rowboat, the *Fox*, made for the crossing (Engvig, 1984).

The first Atlantic row happened more than a hundred years ago. 73 years passed before the next.

The boat carried no mast, sail or rudder but several spare oars. They had a rough row which they slept during the day and rowed at night to keep warm and be on the watch for ships. They encountered storms and capsized once, losing much of their food and equipment. After 55 days at sea, they passed Scilly Isle, and continued on to Le Havre. They rowed more than 3,500 nautical miles and averaged about 60 miles a day. Harboe and Samuelsen won the grand prize and returned to the U.S. and the oyster row. They became world pioneers in long distance rowing in a small open boat.

Today's Traditional Small Craft Association (TSCA) keeps the open boat tradition alive. The organization's many chapters in the U.S. represent all types of older and newer traditional boats. The goal is to promote, protect and present the small craft heritage to the people. The organization arranges annual rows and sails, including a week of Gunkholing for the Sacramento chapter. When members of a local TSCA chapter are out rowing, it is no competition, but everyone, no matter young or old, likes to come in first or row best. Going back a millennium from the TSCA, I would say that the first rowers to win a row on the American continent were the newly arrived Vikings in this new world from Norway via Greenland. Even if we do not know the place or name of the winners, we can say that this would be a natural way to test the new boat made in America. No matter how small, all new boats have to have a trial run.

After John Cabot rediscovered the shores of Newfoundland in 1497 AD, the British established an event that became the oldest recorded form of competition on the continent today. That is the annual row at St. John's. People have been competing in rowing ever since the first British settled long after the Norse had left the northern section of the same island. The official records for the Royal St. John's Regatta on Quidi Vidi Lake dates back to just after the Napoleon Wars, but it is believed to have started long before that. It represents the highlight of the year in St. John's (Beichmann Engvig, 1995). It is a municipal holiday in this part of Canada.

Many different types of crafts are derived from the basic form of the Viking longboat. It has even become an important part of the most well-known old sea shanty and landlubbers drinking song titled, "The Drunken Sailor."

"What shall we do with a drunken sailor... early in the morning?"

And the obvious answer is:

"Put him in the longboat until he's sober... early in the morning. Hoo-ray, and up she rises... "

When sailing vessels developed into heavy carvel-built displacement ships, they often carried one or more open clinker-built boats on deck. The biggest one was the famed longboat and in more modern times, it became the lifeboats on windjammers of iron and steel. The prefix "big" or "long" to the largest boat on board was often used to distinguish it from smaller and handier crafts, which are easier to use when moving the ship in becalm weather, going ashore, doing maintenance over the side, or to collect water etc.

The longboat was any major vessel's best chance for survival if the ship itself should be lost. This type of vessel would also be the ultimate expedition boat and the boat for survival on any voyage or expedition a hundred years ago, as it would have been a thousand years ago. If the Norse or any other ethnic group ventured on the continent in the west they would have used longboats made of wood and iron as a carrier for the expedition.

The Kensington Rune stone was engraved with a language that seems to be Scandinavian, and geo-physicists have now used empirical sciences to establish that the symbols or runes on the stone must have been carved before any settler of the 19th Century could have made them (Hanson, 2001). To institute that its craftsman must have been pre-Columbus is yet another twist to this most controversial Viking or Norse extravaganza in the new world. Could this and other rune stones be keys to open a book of tales from undisclosed boat missions?

The first appearance in modern time York, of longboats sailing all the way from Norway, rowing into the city for the opening of an annual Jorvik Viking Festival to celebrate York's Norse connection. The *Hitra* is first with 18 persons aboard, and then the *Rein*, followed by the Frøya, with the *Ormen* all the way to the left. The fifth vessel, the *Rissa*, foundered the day before at Selby on the way up the river Ouse to York.

THE PERFECT EXPEDITION VESSEL

Our story on the voyages in the *Hitra* is only one of many recorded stories of travels and daring encounters using longboats. It would be easy to make up a long list of journeys in open boats over great distances. Often they came about for the most different purposes, needs or demands. Some of the world's riskiest voyages took place in longboats. There is no reason to believe that such enterprises did not take place even in pre-historic times.

One of the most famed, bitter and resilient journeys in a longboat, is the outstanding sail by Captain William Bligh and his men after the mutiny on the *Bounty*. Many books and several films have been made on the *HMS Bounty*, both on the faith of the crew that obtained the ship and its captain who was forced into the longboat with many men loyal to him. His perilous voyage across the Pacific Ocean in 1789 is an outstanding story of seamanship. He navigated the ship's 23-foot open longboat on a 3,618 nautical miles and 42 days long voyage from a point in the mid-Pacific to East Timor with no loss of life, even when food and water was extremely sparse. All 18 men landed safely on arrival.

Lesser known is the faith of the extreme clipper ship *Hornet* of 1426 tons on a routine voyage from New York to San Francisco in 1866 (Brown, 1974). This big ship of wood had a full load of general cargo from railroad iron and steam engines to barreled kerosene oil. Thirty two persons were on board including two passengers. It made a superb passage around Cape Horn, and after 108 days, had reached equator and the doldrums about a thousand miles west of the Galapagos. That was when disaster struck. The ship caught fire and was soon engulfed in flames. Of the three boats launched, the two smallest were lost after 19 days and nothing was heard from them again, but the 21-feet longboat with captain Mitchell and 15 men including the two young passengers made it ashore. The longboat reached Hawaii after 43 days at sea with all the men close to death. The survivors from the *Hornet* used more time and had a smaller boat than Captain Bligh. They owed their life to the Lord and the small lap-streak longboat of wood and iron according to Mark Twain, who interviewed its captain.

In 1595, the galleon *San Augustine*, which was on a voyage from Manila to Acapulco anchored in Drakes Bay at Point Reyes after three months at sea. On board was traveling clergy and passengers, apart from the ship's officers and crew. While at anchor, a November storm struck and the ship was blown onto the beach and wrecked. A few persons still on board were lost. Fortunately the captain and about 70 men were ashore and so was the ship's longboat. Captain Cermeno managed to get the whole party into the boat and embarked on a voyage for Acapulco more than one thousand nautical miles away. He and his party reached Acapulco seven weeks later (Aker, 1965).

Voyages of this type did not take place only in warm weather or close to equator. Nor did such ventures always result in success. Sir John Franklin's two ships, the *Erebus* and *Terror*, sailed from England in 1845 with 134 men. They were trying to negotiate the Northwest Passage and find a new trading route to the Far East when ice struck the two ships, sending them to the bottom. The crew managed to save the longboats and should have had a good chance of survival. They did not succeed as the participants started to die off. Investigators tend to believe that they ate canned food that was poisoned by lead. What is believed to have been the last observation from the Franklin expedition came from some Eskimos that reported seeing a party of about 40 in the distance dragging a boat along on a sledge. This report tells us that the unknown observed group was probably mariners like the Vikings. The last thing in the world they would leave behind was the boat. Without it, there would be no way to get home. They knew that crossing over any body of water without a boat was impossible.

In almost every church in Norway, there is a ship usually hanging in the center nave. We even find this old tradition in other countries. The main part of the inside of the church itself is called the ship; in Norwegian it is spelled *skip* or *kirkeskipet* (the church's room). The relation between the ship, the voyage and the passing is still as close as it ever was in that maritime nation. The group tugging the boat along on the snow and ice could well have been seamen and final members of the Franklins' Arctic expedition. None of the expedition members survived.

Sir Ernest Shackleton had more luck when his ship the *Endurance* was crushed by ice and sank in the Antarctic in 1915. The 28-men expedition was left on the ice with three small open boats and salvaged provisions. They managed to get to Elephant Island, but knew that it was not a place to survive. With five of his men, Shackleton set out in a 20-foot lap-streak boat to try to sail the storm-swept ocean to South Georgia where there were whaling stations. The 800 nautical miles voyage was a gamble on life and death for the six in the small boat. They endured ice, frostbite, storm and killer waves, but managed to reach the far side of the island after 16 days. The three strongest set out and struggled across glaciers several thousand feet high to get to Husvik harbor. A whale catcher picked up the others. Four new expeditions were launched trying to retrieve the men who stayed behind on Elephant Island. They were rescued after 105 days on that island, living under the two overturned boats left there.

The list of reports on people from lost ships making it to port in the longboat is considerable. In more modern times, and especially during World War II, there were many stories of captains and crews in lifeboats reaching land after long and terrible voyages at sea. The Japanese torpedoed the tanker *Ferncastle* less than a thousand miles west off Australia in 1943. Eighteen crewmembers were killed during the attack or disappeared in the port lifeboat that was never seen again. Captain Andersen and the other eighteen in his boat ended up sailing and drifting clear across the Indian Ocean with very little clothing, food and water (Ljone, 1953). Six perished from injuries, exposure and starvation during the voyage, only thirteen survived the ordeal when they made landfall one month later on Madagascar some 3,000 nautical miles to the west from where their ship had been torpedoed. Completely emaciated and weakened to the point that many had to be carried ashore, they regained their strength and were assigned to new ships.

Some of the bravest incidents were the stories of young lads in small traditional boats from Norway sailing to Shetland during World War II, like Axel Larsen and Hans Larsen in the 22-foot long Aafjords-seksring built in 1899 (Larsen, 1961). Their boat was built on the same farm as the *Hitra*, and it is on display today at the Norwegian Maritime Museum in Oslo. They represent courage and the will to escape from the invaders to fight for freedom. Not only did they barely escape storm and bad weather, they dodged the dangerous German patrol boats and airplanes. Pilots would shoot to sink any boat they found outside the fishing grounds off the coast. This happened often and brave young lives were lost every time it happened. Axel and Hans enlisted in the free Norwegian forces. Axel became a clandestine radio operator in Norway trained in Scotland. Hans was a flight mechanic trained in Canada. Only Aksel returned from the war. Hans Larsen was lost in 1944 when the Sunderland seaplane he was on board disappeared on patrol.

As mentioned earlier, the first of many Viking ship replicas made was the *Viking*, built in 1893. It was sailed from Norway to Chicago, and was the initial one to take on a complete Atlantic crossing in modern times. *Viking* was a full-scale copy of the ship found at Gogstad, built before 900 AD. Captain Magnus Andersen and his seamen performed an outstanding maritime feat, crossing the entire Atlantic Ocean from Bergen in 28 days in the first Viking ship replica ever made. It was sailed to the world exposition in Chicago, The captain later wrote a book about this expedition (Andersen, 1895).

He found *Viking's* construction surprisingly seaworthy and was highly amazed of how well the steering oar worked. He brought a sextant and other maritime equipment and stated that the next sail should try to do without modern instruments. What impressed the Americans the most was the speed of the vessel. A whole fleet of vessels, including military ships, was out on Lake Michigan to greet the Viking ship's arrival. The wind came up strong from behind, and all of a sudden the escorting vessels had a hard time keeping up with this ancient open boat. It got to the point where Captain Andersen had to let up on the sail to help the followers keeping up. *Viking* could easily have been in port well ahead of most of the escort.

In 1980, Swedes reconstructed a small Viking boat and named it *Krampmacken* (Nylen, 1982). It was eight meters long (24 feet), with six oars and a square sail, with the intention to take it on the rivers to Miklagard (Istanbul). They sailed, rowed and towed the boat on water and on land upstream through Poland to the border of Russia, but had to do a detour through Balkan, finally reaching their goal after several years of struggle. This was a reenactment of one of many Viking trade routs to the east. The voyage gave the world insight into the practical details of the Vikings travel on rivers.

Written reports and archaeological finds from the trade with eastern areas have shown that a lot of activities took place along the waterways of Eastern Europe during the Viking Age. Of all the written stories, the Ibn Fadlan reports are the favorites of today, especially the fixation on the stature, body cleaning and sexual activities of the Vikings (Crichton, 1988). To us it shows that the burning of the chieftain's boat was a common sacrifice. This group comprised of well-off merchants who would throw a big party to celebrate the passing on to Valhalla in a good boat. Even if the boat was a principal asset, they could afford it because they simply had more boats.

There is a great difference between this story and the excavations at L'Anse aux Meadows and other places in Scandinavia, where unearthed rivets tell us of social stratification during the Viking Age. What the rich could afford to do did not apply to the general public. Even if the boat was old, worn or even wrecked, parts of it could always be good for something. The average farmer or settler would tend to save and reuse any part of the old boat. The many fragments of iron rivets found at L'Anse aux Meadows confirm this.

Going east was a common thing in the Viking Age, and stories and oral reports handed down must have been plentiful. When the Norse going out of

Newfoundland discovered the same type of navigable rivers in Vinland as in Russia, it would have been the most natural thing in the world to try to sail them and see if another Miklagard or big city was around the bend. Unfortunately, they did not find much gold or silver or Arab merchants to trade with, only Stone Age *Skrælings*. The natural result would be to leave the continent to itself, apart from gathering material to bring home to Greenland.

One definite conclusion can be made from all this: expeditions to unknown lands, rivers and lakes as well as voyages on the high seas, could well been done in what we today would consider small-sized, light weighted lap-streak longboat made from wood and iron and propelled by sail and oars. This is a versatile craft for transportation wherever there is water. The intermediate-size Viking boat seems to be the most suitable vessel for travel and survival on inshore expeditions in unknown waters. Such a boat is small enough to maneuver and big enough to be a good base to serve all the different aspects of navigation discussed here. It is one of the simplest and most versatile boats ever built. All record shows that a riveted and lightly built small open boat can be well-suited for survival even when the odds were slim.

Another original workboat, made in Norway in 1905 for the winter cod fisheries. It served as such up until 1940, fishing out of the Halten Island in the Atlantic. The 22 feet long and 5.3 feet wide *Rein* carries four pair of oars and a square sail. It draws a little more than one foot and weighs about 300 lbs. empty. It can carry four or more people and at least 1/2 ton of equipment without being over-loaded. Picture is taken off Cabrillo Beach, California.

Coming in from the sea outside Antwerp. In gale force wind, I had a full hand managing my old craft. *Hitra* is listing 15–20 degrees. Nobody was out sailing except for the big yacht behind us which did not catch up. When we shot in through the breakers at Breskens, the breakwaters were packed with people. The yacht coming in right behind us was the *Flyer* that had just won the first Whitbread Round the World Race 1978. People were out in force to greet her. Unexpectedly, we showed up in our swift, wet and low riding boat. No wonder we stole the show! Even the captain of the *Flyer* came over to shake hands and have a closer look at the ancient craft they could not overtake.

THE VICTORIAN HERITAGE

IRON DOES NOT FLOAT!

"To make a ship out of iron is impossible"! Common knowledge in the 1800s about different metals and their use associated iron with the sturdiest and heaviest items on any ship. The huge military "ship of the line" would be the vessels with most iron on board. A broadside from one of them was feared on sea and land. Their big cannons were the largest pieces of metal anyone could think of. They were even heavier than anchors that could hold a whole ship in bad weather.

People refused to believe that a whole ship could be made of this material. It was simply beyond their imagination at that time. Ships were constructed with a rather thick hull of a material lighter than water. It was not logical that a much heavier substance could be used in ship's construction. Iron does not float. The common conclusion, even echoed by many scholars was: "the only ship building material capable of floating is wood, period."

When rumors went around that someone in England was constructing ships of iron, it was met with much skepticism and superstition. People on distant shores observed in disbelief at the sight of ships gliding over water, against the forces of nature, with a tall stack that emitted black smoke. Our Lord gave us wind and weather, currents in the sea and everything else. Many good and devoted Christians concluded that the "smoke ship" was the work of the devil, against the law and order of nature, and declared from the bottom of their hearts that they would never set their foot on board such a vessel (Engvig, 1977).

Needless to say, the first iron vessels that were built invoked a lot of mistrust and fear. Despite the fact that iron sinks in water, a few individuals knew that iron could be the ship building material of the future. In 1777, an iron boat, intended for passenger service, was built and launched on the river Foss in Yorkshire. Shortly after experiments with iron plating of a wooden barge for canal transport started. John Wilkinson of Bradley Forge in Staffordshire launched a small iron boat in 1787 and recorded the event in writing. The impossible had been made possible. The advancement could not be stopped. After the turn of the century, the idea to build iron boats for inland transport developed. Engineers who saw the value of the new discovery understood that iron was well-suited to be used as the primary material for future ship building.

But the new invention met many obstacles and problems along the way. It was, for example, a challenge to make a magnet compass work in an iron ship, and without a compass an iron vessel could not take to the sea. The British government formed a committee with the mandate to solve this problem. Early records show that by 1810, a small production of iron vessels for the canals had started in a few places in England (Grantham, 1842). Another bolder innovation was underway parallel to this development.

The steam engines used in boats had its first unsuccessful experiments in France in 1707 and in England in 1736. By 1769, James Watt's machine had been installed in a paddleboat, but the great inventor was indifferent to his industrial engine being modified for use in ship or rail. Work on new types of engines for use in boats continued on both sides of the English Channel. In 1783 a French Marquis tested a boat with a steam engine that functioned on the river Saone north of Lyon, and in 1788 a working steamboat was built in Edinburgh with a Symington engine. It performed well. This experiment was eventually copied in 1801 when the *Charlotte Dundas*, a wooden hull paddle steamer, was built to tow vessels along the Forth and Clyde canals. The ship was a success and did great work until it had to withdraw from service because of complaints about washouts of the banks by the wave action. Four years later, Fulton, an American, and Henry Bell of Glasgow got a steam engine from Boulton & Watt in Birmingham for propulsion of a steam vessel in New York. It was in 1807 that the *Clermont* started her service on the Hudson River doing a speed of 5 knots. Two years later, the *Phoenix* made a sea passage from Hoboken to Philadelphia, and in 1812 Bell's *Comet* started her passenger service from Glasgow to Greenock and Helensburgh. Owing to these successes, steamboats started to appear on many principal rivers. However, all these steam vessels were wooden boats.

By 1821, Charles Manby had taken out a patent in France for an iron paddle steamer, and the ship was built at Horsley Ironworks near Birmingham and named *Aaron Manby* in honor of the constructor's father. It ventured out of the canals of England, making it the first seaworthy steam iron ship. It did trial runs on the Thames in 1822 under the command of Captain Charles Napier. Then the ship loaded a cargo of iron castings and linseed in London for Paris and steamed to Le Havre. After a coaling stop, she would continued up the Seine to Paris for discharging of the cargo. The *Aaron Manby* was later employed on the river Seine. A sister ship was made shortly after by Horsley. Due to regulations, she had to be shipped to France in parts, reassembled, riveted and made ready at Manby's ironworks-shipyard outside Paris. Two more iron paddle steamers were made at Manby's Seine yard. From 1822 until 1830, when the second French revolution took charge of the future fate of these ships, the *Aaron Manby* never required repairs, although she had been sitting on the bottom at low tide even with cargo on board.

Marine engineers followed the development closely and modified and improved the new ship construction steadily. In the late 1820, iron steamers had taken up regular coastal service between several major cities in Britain, and by

1830 the largest, *Princess Royal*, was more than 800 tons. This was less than eight years after the voyage to France by *Aaron Manby*. At about the same time, major ports had started using steam tugboats to tow sailing vessels in and out of port. In 1832, four vessels were being built for The British East India Company. The iron steamship had set out to conquer the world. For the colonies, this development was of the greatest importance. Scheduled sailings to far away countries was important for the mail and packet service, and picked up extra boost from an increasing number of people going to and from the new continents along with the discoveries of gold in new areas of the world. Regular sailings to India via the Red Sea boosted the old idea of digging a shipping canal through the desert at Suez. The next twenty years saw a gradual increase in the building of iron steam ships and their size grew.

By far, the greatest part of all shipping of passengers and goods still went on traditional sailing vessels. These ships got yet another competitor in the iron-built sailing ships. Conversion of some large steamers into sail also helped pave the way for this development. Some of these early steamers whose times had passed became even more famous without their steam propulsion.

These two pictures show the same iron ship. *S/S Pereire* was the pride of the French America Line when built in 1865. For 22 years she steamed across the Atlantic. After conversion into a 4-mast full rigged sailing ship the *Lancing*, she became a real legend doing record breaking passages year after year. At 51 years old, the *Lancing* sailed from Cape Race, Newfoundland to Shetland in an unbelievable 6 days and 18 hours. Her top speed was 20 knots under sail alone, a much higher speed than she ever acquired with her steam engine. Courtesy Norsk Sjofartsmuseum.

The 4-mast full rigged ship *Lancing* (ex *S/S Pereire)* achieved greater speed under sail than she ever did on steam. At forty years old, the *Lancing* established herself as probably the speediest windjammers the world has ever seen. Most important were the facts that the iron ships lasted longer with less maintenance, and they could be built larger than their wooden counterparts. The payload, range, speed and revenue per register-ton for iron ships were far better.

This happened at a time when the increase in cargo volume and crew cost made it more efficient to use bigger ships that could employ a smaller crew relative to its tonnage. An increasing number of iron sailing ships were built after the mid-19th Century. Crew-intensive ships with a small tonnage, such as the brig, could no longer compete and became obsolete. Sailing vessels of wood were gradually forced out of ocean shipping and went into costal trade and local shipping, for example, on the North Sea.

But the U.S. carried on the tradition of building big sailing vessels of wood for ocean shipping until the golden age of iron ship building had culminated. Several factors contributed to this. Although the U.S. was the greatest producer of iron in the world, the nation's production of tonnage for international shipping did not renew itself with the use of iron. It got to a point where America's top leaders were discussing a restoration of its merchant marine, the main explanation being the one mentioned.

While the British were preoccupied with consolidating an empire where the sun never set, and for that reason needed the new inventions in iron and steam to be employed on a worldwide scale, the Americans were busier on the home front developing the new continent and building and expanding the West. Before the railroad to California was completed in 1869, all major cargo shipments to the West Coast had to go around Cape Horn. Inconvenience with discharging and reloading big loads of cargo at Panama made such transport across the Isthmus an undesirable option. For obvious reasons, the New World demanded huge tonnage for its long coastlines including the Great Lakes and rivers. The U.S.'s traditional sailing vessels took care of international shipping and import of exotic overseas products, even participating in world trading in general. Passenger steam vessels of wood and iron were largely employed in coast and river traffic. As late as 1893, apart from the Great Lakes, the British ship building industry still accounted for 82% of the world's output of mercantile tonnage for ocean transport of more than one million tons. By then, the major ship building material was steel.

THE IRON SHIP

Xylograph of the *Great Britain* on the river Thames in the presence of young Queen Victoria.

One ship, more than any other, paved the way for the new development. The construction of the iron steam ship the *Great Britain* was one of the boldest and most advanced enterprises in ship building ever. The building of this grand new creation, a mastodon forerunner of all capital passenger ships of the world, took place at the Great Western Dockyard in Bristol. It was completed in 1843 after about five years of planning and construction. The ship measured 3,443 gross tons and was 320 feet long from figurehead to taffrail with a 289 feet long keel. It was laid down in December 1839. Her beam was 51 feet. The weight of this iron ship alone was almost 2,000 tons.

This ship was constructed of the finest quality Low Moor puddle iron and was fitted with 20 overlapping strakes of plates ranging in thickness from one inch by the keel, to 5/8 and 1/2 inch at the top, except for the upper plate, which was 5/8 inch. She was clincher-built, with double sets of rivets throughout. The ribs and frames were of six-inch angle iron. She was very well-constructed and was given five watertight iron bulkheads, and the decks of wood on sturdy iron beams consisted of a lower cargo deck, two cabin decks and the upper top deck.

The *Great Britain's* propulsion was a four-cylinder steam engine driving a six feet iron screw. She was the first ocean going passenger ship to apply a propeller, and the first to carry six masts with a fore-and-aft rig and a clipper bow. At her launch, she was by far the largest ship in the world. Engineers from all over the world paid great attention to the progress of the construction of the "Mammoth," as she was often nicknamed (Grantham, 1842).

Why did such an undertaking take place at a time when the world had hardly opened its eyes to the new idea of iron built paddle steamers? It is clearly not enough to state that such a vessel was needed. Indeed, her first voyages indicated that she was not. Common sense would also suggest not putting all the eggs into one basket, but rather build more that one ship of lesser size to spread the interest and decrease the risk, and at the same time see how this new screw ship could compete. After all, this happened at a time when the average size ship employed in ocean trading was probably no more than 500 tons.

The British were in a prosperous and optimistic mood after 1814 and the Vienna Congress that would mark the end to all wars. Young Princess Victoria had been crowned in 1838 as Queen of The British Empire. She was fond of gaiety and dancing when she was young. The initial period of the Victorian era was a time of high spirit and optimism. The society had the ability and will to create new and bold innovations, and the persons in power had the courage to follow through. In January of 1845 the *Great Britain* performed sea trials and a trip to London before making her maiden voyage to New York in the summer of the same year. *Great Britain* became the first propeller-driven passenger vessel to cross the Atlantic.

The previous era had been a period of rivalry between the French and the British in the advancement of technology and science. After the first French revolution, many great French scholars from the upper strata of society, often very well educated in mathematics and physics, were forced to flee their country. Old rivalry turned into cooperation when former French Royalists and British engineers and scholars joined forces. It is tempting to conclude that invention and progress got a boost from this symbiosis.

One of these new immigrants to Britain was Marc I. Brunel, a young naval officer, who was well-trained in ship construction. He had a brilliant career in new technical constructions and was knighted. His son, Isambard, with the royalist middle name of Kingdom, grew up in England and inherited the interest for technology from his father. He was educated in Britain and France and became the force behind the construction of the *Great Britain*. But Brunel had a whole team of splendid co-workers from the Great Western Railroad, first and foremost Mr. Thomas Guppy, the talented managing director of the Works.

The *Great Britain* had a very long life at sea and even longer as storage hulk and an abandoned wreck in the Falkland Islands until she was salvaged and towed back to Bristol on a barge in 1970. As mentioned earlier, Karl Kortum of San Francisco was amazed of how well she had kept up when he visited her in the Falklands' in the early 1960s together with Peter Stanford, Ed Zelinsky and Norman Brouwer. He even looked into the possibilities of saving the hulk if the British did not (K. Kortum, personal communication, 1992).

Today the *Great Britain* is on display in her birth cradle the Great Western Dock in Bristol. She is truly the crown jewel of the British maritime heritage with regard to cutting edge design, size, age and originality.

IRON SHIP AFLOAT

Several iron vessels are still steaming on rivers and lakes. But some of these vessels have been repaired and restored excessively to elude or appear as an iron ship even if the hull now is of steel. A few larger original vessels are on land or permanently imbedded on land, like the *Meiji Maru*. Others have ended in shallow waters in far away places and left there. Some of those ships, seem to have survived time and deteriorating forces very well at lonely beaches. Examples are the bark *Lady Elizabeth* in the Falklands, and the *Munzo Gamero* and other iron ships in the Punta Arenas area. Those vessels have been preserved in relatively original states with regard to whatever is left of the ships' iron.

Hulks that have been salvaged to be kept as an historic ship tend to be heavily restored topside and rebuilt with steel. There are examples of vessels that in its appearance of today are close to a complete replica of the named ship. Today, a seaworthy vessel has to meet high standards on safety and satisfy a lot of regulations. Rules are applied to all ships that are intended to carry passengers. New safety requirements make it impossible to keep ships as true historic documents of the past. Such ships will gradually become copies of the original due to new requirements, repairs and renewing. At some point, they will cease to be ships made of iron altogether.

A true historic ship should probably be kept in a safe and protected environment and only be used for special occasions, while a modern replica could be built to give future generations a feeling of how it appeared in the old days. The many wooden replica ships of famous historic vessels from the Gogstad Viking ship and *Mayflower* to the *Lady Washington* and the *Amsterdam* are good examples of presenting ships of the past to modern people and all of them are in good sailing condition.

Old burial ships like the Faro's boat and the Norwegian Viking ship from Oseberg and Gogstad and a few others are the only ancient and more or less complete boats, but they are not seaworthy vessels. Other very old ships kept for interpretation and display are usually restored from wrecks of sunken vessels or salvaged on land, and are often far from being complete. We will continue to concentrate on iron vessels that have been or are still employed in exposition or sailing. The rest of this chapter is an overview of the best-known major iron ships from the high seas that have been saved afloat. They are being preserved, and presented as examples of the first period of ship building in a material other than wood. It is particularly interesting to underscore that the preserved iron ships represents the oldest group of sea-going vessels left in the world.

Of all the ships built of iron, the ironclad steam frigate HMS *Warrior*, built by Thames Iron Works & Shipbuilding Co. at Blackwall for the Royal Navy in 1860, is the biggest and probably also the most original. Only 10–12 small patches less than one meter each have been applied to her original hull over the years (G. Fuller, personal communication, 2001). The *Warrior* is an iron ship even today. This ship was constructed from one inch hull plating produced by the Thames

Ironworks. The ship's measurements are quite impressive. The displacement is 9,210 tons and her length is 389 feet. She was fully rigged as a ship and had a steam engine of 1,250 hp. On her trial run, she made an impressive 17.5 knots. When commissioned, she was much larger and stronger than any ship of her time. She made all other warships in the world obsolete overnight.

Several other ironclads, monitors and gunboats exist in different parts of the world and in different state of protection or deterioration. Some have been kept afloat and are on display.

It is doubtful that any regular iron-built ocean steamer of any considerable size exists today in a relatively original and seaworthy condition. It is likely that the small coastal research steamer the *Hansteen* is the one that comes closest. Not surprisingly, the cargo carrying sailing vessels seem to have survived in greater number than contemporary iron steamers. The obvious reason is that a steamship always represented so much more value even as a wreck compared to a sailing hull. The latter would often be left in remote places of the world to serve different purposes, rather than being towed all the way to the nearest breaker at a considerable cost. It is from this remote collection of hulks that some of the oldest existing iron sailing vessels have been collected to undergo conversions to bring them back to their former glory. The *Great Britain* is a classic example.

The iron bark *James Craig* at Sydney, Australia was built in 1874 by Bartram, Haswell & Co, Sunderland, England at 671 gross tons and a length of 179,5 feet. This vessel was found as a hulk in Tasmania, salvaged and taken to a pontoon in Darling Harbor in 1985 for repairs. The goal included "fitting of new frames, plating and other metal work." Repairs were extensive and major parts of her hull were replaced with new material. This ship was even fitted with a brand new diesel engine that she never had before. *James Craig* started her new life as an auxiliary bark, and made a nice impression, as all her new steel plates and angels had been riveted.

A similar restoration scenario was experienced for a couple of other hulks that have been brought back to life as restored and renewed versions of the old vessels.

Bark *Elissa*, now at Galveston, Texas was built of iron in 1877 by Alexander Hall & Co in Aberdeen, Scotland. She is 489 tons and measures 170.5 feet long. After being found at a breaker's yard in Piraeus in the 1970s as a beaten-up old Mediterranean motor-freighter, she underwent preliminary repairs and was towed to the U.S., where she was completely restored. Bark *Elissa* appears today with parts of her hull made of welded steel and other parts made of riveted steel, but she is considered an iron ship. This was an early restoration project which attempted to bring a ship back to sailing condition. Today, *Elissa* still has a motor, is seaworthy and in good condition, and is an active sailing ship (S. Hyman, personal communication, 2001).

Falls of Clyde in Honolulu, Hawaii, is the only four-mast full rigged ship left. Russell & Co in Port Glasgow, Scotland, built her in 1878. This ship is 1809 gross tons and measures 266 feet long. She has seen repairs but is to a large degree, intact from her working days and is a fine example of the great sailing ships built of iron from the later part of the iron ship building era in Great Britain. She is a stationary museum ship and floating exhibit at the Hawaii Maritime Center, Honolulu.

To maintain the wood in this large four-mast iron ship in the climate on Hawaii is a major challenge.

The iron bark *Polly Woodside* in Melbourne, Australia was built in 1885 by Workman, Clark & Co in Belfast, Northern Ireland. She is 678 tons and 186.5 feet long. She was saved in the 1960s at the end of her working career and taken to Melbourne. The subsequent extensive and painstaking restoration has brought her back to an iron bark, as close to original condition as possible. This ship has received an award from the World Ship Trust for authentic restoration and rigging. *Polly Woodside* is a floating exhibit on the River Yarra and the centerpiece of Melbourne Maritime Museum (Brouwer, 1999).

The large full-rigged ship the *Wavertree* in New York, built in 1885 by Oswald-Mordaunt & Co, Southampton, England is 2,170 tons and 279 feet long. She was acquired by South Street Seaport Museum from Argentina after her days as a sand barge had ended in 1970 and towed to New York. Her iron hull has been repaired and some doublings have been added. She has undergone extensive restoration and renewing topside with new masts, yards and rigging. This has given the ship its original appearance once more. Her size is impressive. The *Wavertree* is still to a great extent an iron vessel. She is the biggest cargo handling wind ship of iron still afloat (T. Miles, personal communication, 2002).

Af Chapman is a full-rigged iron ship stationed in the City of Stockholm. She was built by Whitehaven Shipbuilding Co. at Whitehaven, England in 1888 as *Dunboyne*. This ship is 1,428 gross ton and measures 256.9 feet long. She sailed as a world trader with Irish, Norwegian and Swedish owners until 1915. The Royal Swedish Navy bought her for cadet training in 1923 and renamed her. In 1947 she was sold to the city of Stockholm. From 1949 until today she has been a stationary floating youth hostel in Stockholm. Her number of overnight stays has passed one million. Ongoing maintenance and the fact that she never really went out of business have kept this ship in good shape. *Af Chapman* was built when most of the established shipyards in Great Britain had converted to steel ship building. She is a late-built iron ship, and might well be the youngest of all the iron-built windjammers.

Stockholm's *Af Chapman* built in England in 1888 may well be the youngest iron built tall ship left in the world and could also prove to be one of the most original iron hull windjammers anywhere.

The full rigged former Danish sail training ship *Georg Stage*, today's *Joseph Conrad* of Mystic Seaport, Connecticut was built of iron by Burmester & Wain, Copenhagen in 1882, and fitted with an engine. With her 203 tons and measuring 118 feet, she is a small ship, and in good condition. She has received a modern laminar protection coating outside and is cemented on the inside under the waterline down. Like many other iron ships, she has been cemented inside for ballast and to protect the iron. She seems to be the only floating sailing vessel of iron left that is not built by the British. *Joseph Conrad* is a stationary part of Mystic Seaport's great collection of ships and maritime artifacts.

The iron bark *Star of India* in San Diego is by far the oldest ocean trader still seaworthy and also the most complete. Her iron hull is still intact as she was built in 1863, but she has a steel doubling along the waterline. Occasionally she is taken out for a short spell under her own canvas. A report on the research done on her iron can be found in the last chapter of this book.

It is beyond my intention to continue listing former military and other, mostly small iron boats. The overall conclusion is that there are few vessels of iron left in the world that still carry the major part of the iron hull from their construction during the Victorian Era. Those few that exist will serve as a basis for research for generations to come. It is interesting to note that while U.S. and Australia never had much of an industry for building tall ships of iron for world wide service, seven out of the nine old windjammers presented in this chapter have found their homes in those two countries.

BRITISH IRON

The British Isles became the focal point of iron ship building from its slow start through its golden era up until the time when steel ships gradually replaced iron ships. Key ecological, human, political and societal factors facilitated the transition from wood building to iron construction. Many of these same factors mirrored processes important during the Viking Age with regard to the development in ship design and construction.

After the first Vikings landed at Lindisfarne, the East-men from the North started on a mission to conquer the Northern Ocean. The Vikings who established the Kingdom of Norway had good fuel and good quality iron, abundant sources of material for ship building, and lots of experience with boats, sailing and navigation. It was a basic necessity in a rugged country where most of the people lived by the long fjords, on islands, by the bay, or along other parts of the coast. With mountainous land almost everywhere, and with dense forests full of wild animals, the boat was the main means of transportation. They needed lightly-built vessels that could be beached and protected from wind, sun and snow. This society needed a large number of boats and ships, for transport, for war or for crossing the sea. The boat design they developed is still with us today.

When the British, with their vast amounts of coal and poor quality iron ore, had developed new iron producing methods to utilize these resources, the future was set for a development that would revolutionize transportation. This development, however, took time. In the 17th and 18th Century, the production techniques for iron gradually improved.

Sweden had exceptionally clean iron ore and good firewood. It became a major producer of iron and iron products for much of Europe. The Lancashire method (see the laboratory reports in the following chapter) for the production of wrought iron became popular in Sweden due to the low grade of sulfur in coal made of wood. Ironworks in Norway also produced excellent wrought iron, and Fritzøe (see later) exported bar iron to America under its stamp mark. It was commonly referred to as "Norway Iron" by the Americans, and became so attractive that the name was employed even for Swedish iron sold in the U.S. (Espelund, 1995). Complaints about this ended on the King's table in 1867, but the King deferred it. The quantity of Norwegian produced iron, however, was always small compared to Swedish iron.

Great Britain imported a lot of iron from Sweden until Henry Cort in Fareham introduced a new technique for eliminating sulfur from the poor quality British iron ore in 1784 by keeping the liquid iron separated from the sulfur rich heat from burning coal in a reverbatory furnace. During this operation, the liquid iron was puddled. This production method became known as the Puddling process. The wrought iron was also called hammer iron. Further work on the product required hammering. This was extremely strenuous work done by men. Water driven hammers were employed until the introduction of steam hammers. The hammering was an essential part of the refining process. Mr. Cort also patented grooved rolls to roll out the impurities as an aid for better production. The name "hammer iron" rather than wrought iron was adapted in some countries for good quality iron for ship building.

Mr. Cort's process for more efficient production of better quality iron became the basis for production of iron all over Great Britain. The techniques were improved several times over the following period and adapted to different local types of ore. Best Yorkshire was established as a brand name for superior quality. Only a few ironworks on the British Isles could use that label or brand name. Low Moor Iron Works near Bradford in Yorkshire was one of them. It delivered the iron for the construction of the *Great Britain* and also some of the iron for the *Hansteen*. The ironworks producing Best Yorkshire quality could ask better prices and Low Moor iron was known to be expensive.

In the 1830s, Joseph Hall, the "Iron Man from Tipton," pioneered a new process for making better iron at Bloomfield Iron Works in Staffordshire. This superior quality iron was known as BBH and was marketed under that brand name. Plates in the *Hansteen* still carry this original stamp mark, as described in detail in later chapters. Bloomfield was among the ironworks outside Yorkshire that aimed to produce Best Yorkshire quality iron. To a large extent, it did, and became well-known around the world for its good ship quality iron (Griffiths, 1873). Their product was graded into B, BB and BBB for best, double and triple best quality. The Bloomfield stamped iron in the *Hansteen* is labeled BB and BBH BLOOMFIELD; the last mark is assumed to represent Bradley, Barrow & Hall, the principal owners of the ironworks.

The salvage tug *Oscarsborg*, built in 1874 by Akers Mek Verksted, Oslo, was constructed of Bloomfield iron according to shipyard records from the ship's building. Observation during docking has shown that this tugboat still has a riveted hull. It is believed that this mainly consist of wrought iron plates. Only future investigations will determine if this is the case. By now, this vessel has seen 127 years of professional service as a tugboat. This is an extremely long time of active duty for any ship and indicates that its hull must have been constructed of a sturdy and lasting material. Both the two yards in Oslo constructing iron ships back then imported building material from Bloomfield in England. Their two surviving ships mentioned here were both constructed using their iron. It shows that the Bloomfield iron must have been of a very lasting quality.

A survey shows that the plates under the waterline of tug *Oscarsborg* are riveted. Only future investigation can tell if they still are the original Bloomfield plating laid down in 1874. Photo August 2004.

The area around Birmingham was a region rich in coal and iron ore. It was named the Black Country for its many coalmines and ironworks producing iron from local resources. The district also had established a good system of canals for transport in and out of the area. By 1830, the Birmingham-London canal transported more than a thousand tons a day. Eight years later, the same two cities had been connected by railroad. Wednesbury, where the plates for the *Star of India* were made, was only a few miles from Tipton and less than 10 miles northwest of Birmingham. *Aaron Manby* was built at Tipton and it was the home of Joseph Hall, which was producing better iron at Bloomfield. The Black Country was not the only iron district outside Yorkshire. Mines, mills and shipyards had popped up all over the British Isles before the mid-19th Century.

Lloyds of London, the old and distinguished ship classification bureau, established new specifications for the construction of wrought iron vessels in 1855. It determined a required minimum strength for all iron material to be used in a ship. The goal for iron works delivering ship quality iron was, of course, to be well above this demand, as it was for any serious shipyards to build good quality tonnage. The basic material in all iron ship building was different sizes T - and angle iron, flat iron with a bulb and iron plates in different thickness. Those few profiles, developed between 150 and 200 years ago, has been supplemented with some new profiles today. But in general, the bulk of the steel in any shipyard today do not look much different than the iron that was produced back in the 1850s. However, when it comes to purity, quality or strength the story is very different.

Lloyds register statistics of the gross tonnage of iron and steel ship built under the bureaus survey clearly shows the development in the use of iron and later steel in the ship building industry. In the year 1877, 443,000 tons of iron ships were made and no steel ships were made. Three years later, in 1880, 460,000 tons of iron ships and 35,000 tons of steel ships were made. Ten years later in 1890, the shipyards produced 42,000 tons of iron ships and 1,000,000 tons of steel ships. This development continues. Production of iron tonnage had dropped to less than 8,000 tons five years later. After the year 1900, no iron ships were made, but the steel tonnage had increased to 1.5 million gross tons. These numbers shed light in the general development in ship construction, but the numbers must be adjusted for tonnage built outside the bureau and in other parts of the world. Even so, major ship building nations show a similar pattern.

Producing iron has always been a difficult process that demanded good craftsmanship. Modern experiments with making bloomery iron from bog iron ore has proved that the artisan ironworker in the Viking Age must have been well-trained and in command of a lot of special knowledge. The iron makers also needed high quality raw materials to work with to make a first class product. Yet the use of iron was so common that iron production must have been an important part of the life in the Viking society, just like woodworking and building of ships must have been.

To build an iron ship was simple and complicated at the same time, and it required more time and several extra work operations than needed to weld together a steel hulled vessel like the way we do it today. Angle irons were heated in big ovens until they turned glowing white. Then they were pulled by men to the framing board and given the right curve and angle. This work had to be done quickly, while the angle was still red hot, with all the men working as a team. The job was hard, under hot conditions and often done under an open roof. Slowly, frames and some strikes of plates were screwed together to form the skeleton of an iron ship. When the outer streaks were on top of the inner ones, riveting started. Free and readied holes were riveted by two riveters hammering on the white-hot rivet.

A riveting team consisted of five men. Two were needed on the outside when riveting was done manually, while three were needed inside. During the restoration of the *Hansteen,* we riveted some 3/4-inch rivets in the keel by hand hammering. We documented the operation on film. It was interesting to learn that when a compressed air hammer was used, the time taken to complete the task was about the same as when riveters did it manually. But hand riveting was far more physically strenuous, and one more man was needed.

Today, when restoring an iron built ship with modern machines and compressed air tools for broaching, countersinking and riveting, it is forgotten that the iron ships of the mid-19th Century were completely handmade out on the slipway with the use of hard labor and hand-held tools only.

Wilhelm Trygg and Gunnar Pettersen hand riveting the old way, while being filmed. The upper row of holes is for later riveting on of the keel shoe that was completely worn out. No one could forge the new shoe, so we had to go with an L-profile with a welded-on flat iron for the new riveted shoe.

Above: A few of *Hansteen's* bottom plates were replaced. Right: Inside the engine room, holes readied for riveting.

Riveter Wilhelm Trygg at Nyland redoing some "lost" rivets in old iron plates 115 years after the yard's "fathers" riveted *Hansteen* together. For a greater part of the original hull, this was the only thing the government's ship surveyor demanded so that the ship is considered seaworthy. The new plate underneath the old ones and abaft is a second replacement with steel where the original iron had been damaged during the ship's sailing years. Notice the air tools — the Ingersol-Rand machine in front and the Atlas nut puller.

The last operation building or restoring a riveted ship is the *dikting* that guarantees the hull will be watertight. The chalk applications of the wavelike marks on the bottom of the *Hansteen* show that this streak is ready for sea. Wilhelm Trygg at work on the *Hansteen* at Nyland in 1982.

BEING PART OF HISTORY

Knowing the story of the old wrought iron ships that are still with us, it is difficult not to be impressed with the position Great Britain has played in this part of history. Substantial production of iron and even the building of iron ships did take place in other countries. Scandinavia is already mentioned, but Germany, France, Russia, the U.S., as well as other parts of the world developed heavy industries based on the production of iron. But in the production of sailing vessels, no one in the world could beat the British. They turned out hundreds of big sailing ships of iron after the mid-19th Century. Over a period of not much longer than 30 years, they provided the world with what was needed for cargo shipment for years to come. No one could compete in this mass production of "inexpensive" windjammers.

For ocean steam ships, it is very much the same scenario. The sad part is that of all the proud old iron steamships for ocean transport, none seems to have remained in seaworthy condition. Culmination of the construction of big ocean steamers of iron occurred in 1880 with the building of the Inman Line's *City of Rome*, a proud and beautiful steamer, 560 feet long, with three smokestacks and four masts. When she was launched in 1881, she came second in size only to the *Great Eastern*, Brunel's greatest iron ship that never saw success as a liner. For iron-built ocean steamers in passenger service, the *City of Rome* was the last. Bigger and finer passenger ships were constructed as time passed on but they would all be built from the new and better material — steel — as the *City of New York* and *City of Paris* in 1887 (Johnston, 2000).

For sailing vessels and smaller steamers, iron ships were still being built for some years to come, but in sharp decreasing numbers. It was also an option to finish off the old material, as classification records sometimes state, by the use of iron and steel (Det Norske Veritas, 1892).

The small steam schooner *Vaerdalen*, built in 1891, is a prototype in iron and steel based on the U.S. lumber schooners made of wood. Det Norske Veritas approved the construction. The belief is that her hull consists of iron plates. The ship's bottom has never seen repairs nor has it been analyzed. The *Vaerdalen* has seen 85 years of continuous service as a fjord steamer for post passengers and cargo, and as a cargo freighter on the coast of Norway. It is undergoing a painstaking slow restoration.

We have named a handful ships from the seven seas. They are among the oldest vessels still afloat and part of a long-lost history of iron in ship building.

Two ships in particular will be discussed in the following chapters. The bigger one was built in Great Britain, the smaller one in Norway. Research done on the iron and the stamp marks of the bark *Star of India* and the sail steam ship *Hansteen* will be highlighted through the unpublished reports from these investigations. Both vessels are from the 1860s, and they have a long and exciting history. Furthermore they are more or less directly connected to the era of new development and technology.

The Bloomfield Iron Works, which used the coal from the Bloomfield Coal Mines to produce iron that are still present in the *Hansteen*, applied James Watt's steam engine no. 1 in its mines for many years. The James Watt steam engine is regarded as the main invention that triggered the industrial revolution. The stamp marks on the *Hansteen* iron also carry the signature letter of Joseph Hall, "the iron man from Tipton," that greatly improved the production of iron.

The *Star of India* could be connected to George Stephenson's locomotive, another crucial invention in this technical revolution. Not only did Director Losh at the Walker Iron Works provided financial aid in the making of George Stephenson's first locomotives, it seems he participated in the construction as several original patented drawings carry his name. William Losh stamp marks are still all over the frames and contra angels on board the *Star of India*.

Research led me to this hulk, named the "Old Boat" by the homeless people of Oslo who occupied it for many years. It was in fact Nylands hull # 11, constructed in 1866 only a few yards from where this picture was taken in 1977. Named *Hansteen,* this vessel was used to transport King Oscar II of Sweden and Norway in 1872. When I investigated her lower hull, I found almost all her interior wood and cabinets intact. No readings on the ultrasound meter showed plates measuring less than 7 mm of iron, which is a valid number in a 100 feet and 120 ton ship.

HAMMER IRON AND QUALITY MARKS IN PRESERVED 19th CENTURY SHIPS

Practical tests of the wrought iron rivets in old lapstreak construction in open boats of wood during the voyages described earlier in this book taught us a lot about the iron. The boats have been given the ultimate test of buffing wave action, in shallow as well as in deep water, and choppy sea. Not even a single iron rivet gave in during our voyages — all of them are still in place, where the boat builder at Haarstad in Aafjord installed them back in 1863. This makes it impossible to examine the boat's iron rivets under the microscope in a laboratory as I am not prepared to sacrifice any rivet for that purpose.

The verdict must be that "farm iron" from the 19th Century was of extremely good and lasting quality, which could withstand salt water very well. In comparison, what has research and testing of industrial iron in bigger ships, made during the same period, reveal? Ships made of iron from the same time are still around. Examining the iron in some of these ships will give us an indication of its durability compared to the farmer's iron in the open longboat.

The rest of this book will differ from the previous chapters. It will be based on the two reports on research done on the wrought iron in the sail steam ship *Hansteen* and the bark *Star of India*. These two unpublished papers were accepted for presentation at the 6th Maritime Heritage Conference and the 4th International Ship Preservation Conference, at the historic seaport of Wilmington, North Carolina, 25–28 October 2001.

Not everyone reading this book wants to delve into all the details on the investigation and results of the research included in these reports. This chapter is an introduction and a presentation of the findings documented in those two reports. Laboratory reports from A/S Norsk Jernverk at Mo, included in the *Hansteen* report, are all written in Norwegian and contain drawings, diagrams and pictures. A register of footnotes and references included in my original reports are omitted here. This has been a difficult decision to make, but had to be done, as the book has relied on another presentation of the reference sources.

Submitted to the same conference in North Carolina was also an abstract with the title: "The First Shipyard in America," which was a brief presentation of the material presented in the first part of this book, regarding the Viking's ship building with iron rivets, and my research on the construction and use of traditional open boats of wood. This presentation has since been published in Viking Heritage Magazine (4/2001).

The following is a summary of the reports on wrought iron and quality marks in the two named ships built shortly after the mid 19th Century. More detailed reports can be found in the next two chapters.

The first report was written during the restoration of sail steamship *Hansteen* between 1979 and 1993. The *Hansteen* was built by Nyland in 1866 for the Government of Norway as a depth sounding and mapmaking vessel. It was 120 tons, 101 feet long, with a beam of 15 feet, and was fitted with a two-cylinder, high-pressure steam engine and six sails. From 1867–1897 the ship made depth soundings off the coast of Norway. From 1899–1948, it was a local passenger ferry. Before being laid up, she had 81 years of service on the North Atlantic coast of Norway. From 1949–1977, she was a lodging facility. When phased out of this service, she was scheduled to be burnt when restoration was initiated. In 1986, the hull and exterior of the ship was brought back to its original appearance, using old riveting techniques. By 1994, the *Hansteen* had been completely restored as a surveying vessel and to an excellent and seaworthy condition. This work earned the restoration team national and international acclamation and the first Gold Medal of Merit ever given to ship preservation and maritime heritage from the King of Norway.

Fritzøe Iron Work's rolling mill at work, as depicted in the Norwegian Skillings Magazine, 1859.

The shipyard's records from 1866 showed that hull # 11, the *Hansteen*, was constructed of iron from Bloomfield and Low Moor in England, from unknown ironworks in Sweden and Fritzøe Iron Works in Norway. Iron stamps found on board this ship on plates and angle iron are marked with impressions from one of the ironworks, namely Bloomfield. Laboratory tests done on 27 samples from the hull confirm the records. My report revealed that even after complete restoration in 1993, *S/S Hansteen* was floating mainly on her original iron hull from 1866. She was still a wrought iron vessel then, and one of the few left. Only seven iron plates that had been physically damaged during her last 30 years were renewed with steel during the restoration. The ship had probably been riding on the bottom, eating away the plates from the outside, during low tide in one of its hostel harbors. When we removed the concrete we discovered that she had been floating on this material rather than on iron in a few places.

Stamps made by the iron manufacturer for quality and identification give proof of originality and are important identifiers for the iron in any old ship for several reasons: they open up for future research into the ironworks and ship quality iron; and tell us how the iron was produced, shipped and used during construction. They also help us understand how the iron lasted on board the finished vessel over the lifespan of the ship. Furthermore, markings serve to identify a ship. They give proof of how much of the surviving ship is original, indicate the quality of material used in her construction, and give us knowledge about service and maintenance over a lifespan in an environment hostile to iron. Most ships back then lasted fewer than 30 years. Ships mentioned in this book are all much more than a hundred years old, which are extremely old for any seagoing vessel. Did the quality of the iron and the way the yard made the ship contribute to the survival?

Between the introductions of iron ship building in the first part of the 19th Century and the abandoning of iron during the 1880s, thousands of iron ships were built. Few relatively complete iron hulls in seaworthy condition still exist. Former iron ships have been restored excessively and appear today with much steel in their hull. These ships are lost as genuine historical documents, as specimens for future analysis on the iron, and for research on the shipyards and the original constructional methods.

For commercial ship building, this building material has not been in use since. Old ironworks were closed down, dismantled or converted into steel mills, most of them before the start of the 20th Century. Better quality steel was produced in great quantity by steel mills at competitive prices. Ship owners preferred the new material for future tonnage. The last existing iron ships and the iron material left in them, represent relics that should be saved and protected. These ships can never be restored once the iron has been replaced or damaged.

Hansteen is a seaworthy ship from the mid-19th Century. This vessel is a fine example of early ship building, with many opportunities for research and scientific studies to extend knowledge of this first period of construction on an international scale. It is extremely important to preserve this historical vessel when it was brought back to its original appearance of the 1800s. Original wrought iron structures from the classical era of the 1860s are rare, and ocean-going and coastal vessels with sails and a steam engine, still in service, are among the rarest.

Schooner rigged steam ship *Hansteen* sail trails on the fjord after restoration from a hulk.

The bark *Star of India* is the former full-rigged ship *Euterpe* of 1197 tons, built at Ramsey, Isle of Man, in 1863, for general trade on all oceans. The ship first worked in the India trade, then in the emigrant trade on Australia and New Zealand under British flag. In 1898, she was transferred to American owners for the lumber trade from the Pacific North West to Australia and finally in the salmon fisheries for the Alaska Packers' until 1923. By this time, she had 60 years of service on the high seas. Since 1927, she became a museum ship in San Diego, California.

The main objective of the investigation was to find stamps made by the iron manufacturing works for identification, quality and originality reference. Identification of iron markings provides a platform for further research and serves to identify the ship itself. The goal is to be able to show proof of originality, and indicate the quality of her building and further understanding about the ship's care and maintenance over this extremely long lifetime in a hostile environment that deteriorates iron. Sail ships of the 19th Century usually lasted only a few years. A small percentage stayed in service for 30 years or more. What makes some ships survive for more than one hundred years?

My secondary goal was to try to find impressions similar to the ones I found when doing research aboard the Norwegian wrought iron ship, the *Hansteen*, also a seaworthy vessel, built in Norway with iron from England, Sweden and Norway only three years after the *Star of India*. This would contribute to a comparison of ship quality iron from ironworks in these countries. Scandinavia and Great Britain used dissimilar quality iron ore as well as different production material and methods in wrought iron production.

These countries have an old iron and ship building tradition that dates back to the Viking Age and before. They produced and used iron for an iron-ship building industry in the 19th Century when this type of ship construction became an alternative to ships made of wood. Before the end of the same century, iron ship building was abandoned. Of the thousands of iron ship built between 1840 and 1890, only few seaworthy and original windjammers are left. These last survivors are important. Once iron is replaced with modern steel, much of the ship's historic and original value is lost.

The yard's ordering lists for the iron used to build the ship *Euterpe* in 1863, now the *Star of India*, have not been located and could be destroyed. My survey shows that the ship's frames or ribs, beams and angle iron are all marked with LW & WALKER and CONSETT, telling us that this ship was made from those particular brands of hammered wrought iron (D. MacCallum, personal communication, 2000). It concludes that the *Star of India* is an authentic vessel with the original iron from her building still intact.

This investigation tells us that the bark *Star of India* is not only the oldest existing iron-build merchant vessel left in the world that was licensed for trade on all oceans, but most importantly, she is very close if not one hundred percent original. Her present-day wrought iron hull is the original one, and has few repairs or alterations. This makes her an amazing and unique relic among all the existing historic ships. She even gets under way by her own sails as always.

Bark *Star of India* under sail in San Diego Bay. Courtesy Maritime Museum of San Diego.

This is a true and genuine sailing ship from the mid-19th Century that is well taken care of. I am happy to have been able to help establish that the ship's iron still is where workers placed it when they built *Euterpe* in 1863.

It shows that good quality wrought iron has better durability than any type of steel that became the preferred material towards the end of the 19th Century. It is understandable that ship-owners switched to steel when they learned that a steel ship could take much more punishment than an iron ship. There is no doubt that the new material was better for use on a ship that had a calculated lifespan of not more than 20 to 30 years at the most. These vessels were often written down and paid off in half that time or less. The economic side of the business demanded steel and so did classification bureaus and insurance companies after a while. Like a much later "farewell to steam," the shipping business experienced a "farewell to iron ships."

They were sold for scrap and recycled. After World War I, the rest of the iron built ships were broken up. Interest in saving old ships only lingered with some old seamen and a few others who never managed to raise much publicity on the issue. Where should the money come from? Iron and steel have always been associated with profit. Old iron is scrap. In the 1920s, as in the year 2000, the opinion on old iron is the same. It is best sold to a scrap dealer. Why bother salvaging anything made of iron? Good and well-built iron ships were not kept even if they were floating and doing well. Owners knew that their topside wood sooner or later would start leaking and rot away, letting rainwater into the superstructure and the hold, which led to a drop in value. So they tried to get rid of the ship instead of paying laid-up fees or initiate expensive repairs.

The best example of the resilience of an iron ship is the *Lady Elizabeth* in the Falklands. This 1,208-ton iron ship built in Sunderland in 1879 was abandoned in Port Stanley in 1913. Since 1936, she has been sitting on a sandbar in a cove left to herself with her bowsprit and all the three lower masts standing and the heavy main yard aloft as always. That the yard and masts did not collapse after almost 70 years is unbelievable. She has more or less been left untouched. Several countries have sent expeditions to the Falklands to see if they could save and restore this ship. But some people in Port Stanley do not like to see her go. She is an excellent specimen for research on iron in ship building.

It takes time to educate people in the benefits of saving old iron. Insurance companies, people in shipping and shipyards workers, even ship preservers cannot see why one should keep old iron. More regrettable is the fact that no scientific institution has looked into the questions around the quality of hammered iron compared to steel with regard to exposure, aging and long-time endurance. I have searched extensively to find materials for this book — information on iron in ship building, and how well these old ship lasted compared to newer tonnage. Unfortunately, very little information is available.

Apparently, many people regard iron as just an old rusty metal, which is not worth writing about.

STAMP MARKS AND LABORATORY TESTS ON THE HULL OF *S/S HANSTEEN*

INTRODUCTION

Existing records, such as the ordering books for the purchase of iron, for Nyland Shipyard in Oslo, which built the *Hansteen* as hull # 11 in 1866, show that this ship was constructed of Bloomfield and Low Moor iron from England, unidentified iron from Sweden and iron from Fritzøe Iron Works in Norway (Engvig, 1986).

Iron stamps still found the plates and angle iron on board this ship carry impressions from one of the ironworks, namely Bloomfield in England. Laboratory tests conducted on 27 samples from the hull taken during the restoration period supported the records. The tests confirmed that the lower part of the hull was assembled using Swedish plates, while the upper plates and frames came from Bloomfield. So did the iron for the bulkheads.

Luckily, a lot of records for this ship still exist — not only is an important part of the building archives preserved, but damage reports, surveyor's records and court documents from accidents the ship encountered are also available for study. Since the ship was Norway's first research and surveying vessel with a new design, it received lots of attention from magazines and papers. Even scientists like G. O. Sars (one of the pioneers in ocean research) working on board the *Hansteen* managed to include the ships comfort in his scientific report for 1872. The coastal fisheries inspector at Aalesund suggested the *Hansteen* used as a model for a fishing patrol vessel late in the ships surveying period.

The material in the archives and all the testing revealed that *S/S Hansteen* was still, in major parts, an original iron ship from 1866 when restoring finished in 1993. When the ship had completed its full restoration to steaming and seaworthy condition, only seven original iron plates below deck, damaged between 1948 and 1979, were removed and renewed with steel during the restoration, while 19 steel plates from earlier repairs all had to be renewed. *Hansteen* was in steaming service on the coast of Norway for 81 years before she became a lodging facility in 1948 (Nyrud, 1982; Johnsen, 1984).

Stamps made by the iron manufacturer for quality and identification provide proof of originality and are important identifiers for the iron in any old ship for several reasons.

They give a benchmark for future research into the iron works and ship quality iron, and also in what way the iron was produced, shipped and used during the construction, and how it lasted on board the finished vessel over the lifespan of the ship.

Furthermore, they serve as an identification of the ship itself. They tell of the ship's service and maintenance over a long time, sometimes even more than one hundred years. This is an extremely long lifespan for any seagoing vessel. Why does this material last so well and make a ship's hull survive for such a long time?

CONSTRUCTION AND HISTORIC RUNDOWN

The sail steamship *Hansteen* was built as hull # 11 by Nyland Værksted, Christiania (Oslo) in 1866 for the Government of Norway to be used by *Norges Geografiske Oppmålings Hydrografiske kontor* as a depth sounding and mapping vessel, and if possible, in combination with oceanographic research. It was 114 tons, 101.6 feet long, with a beam of 15 feet, and was equipped with a two-cylinder, high-pressure steam engine (boiler) of 25 NHP and fitted with a screw (propeller). The ship was given full sailing capacity and was rigged as a schooner with six sails including a square sail. The sailing area was 210 square meters. A crew of eight navy AB seamen (professional sailors) was used, telling us that in the days of sail, a steam engine was not entirely trusted.

The hull consists of six strikes of plates and 48 frames. The six streaks were labeled from A to F, with A as the top and thinnest one, partly above deck, and F being the heaviest one riveted on to the keel. The records from Nyland show that the A, B and C streaks consist of Bloomfield iron, while the three lowest (D, E, F) were produced using Swedish iron. All the angles for frames, ice-enforcement in the bow, beams and support for the bulkheads were mainly of Bloomfield BBH Best Best iron, as were the five bulkheads, the coal boxes and the funnel. The boiler was produced mainly by Low Moor, Fritzøe and Swedish iron. It is believed that scepters, stanchions, chain plates, gangway for the rowboats (ship landings like piers were almost nonexistent along the coast in those days) and other forged items were produced using Fritzøe or Swedish iron. Many of these original items still exist. Fox Head & Co was also named as provider of material for ship # 11.

Calculations and records from the trial run also exist. The block coefficient of *Hansteen* is 0.41. That tells us that the hull was made to be easy-driven, probably due to her rather primitive engine and a boiler that was fed with salt water. The boiler was replaced after few years and the engine got compounded in 1879. A new trial run shows that the ship managed 9.9 knots. The top speed would probably be higher with sails if the propeller was allowed to turn along and not break the speed. The Royal Norwegian Navy manned the ship and its captains were always lieutenants in the RNoN appointed by the King (even on *Hansteen's* first trial runs after full restoration in 1993, a lieutenant was in command of the ship, keeping up a 100-year-old tradition).

From 1867–1897 the ship performed ocean surveys outside the coast of Norway. It usually worked out to 10 or 20 miles off shore mapping the bottom, but went even further out to do the rich fishing banks off the coast. The groundbreaking work of mapping the whole coast of Norway took 30 years and earned the navy officers aboard the *Hansteen* international recognition.

Research done and maps made on board the ship was presented at the 1875 World Exposition in Paris and won a gold medal and a diploma for excellent new scientific work. Earlier on, *Hansteen* was recognized for ocean exploration.

In 1872, the ship was taken out of service for 14 days to be on call to accommodate the new King Oscar II of Sweden and Norway during the nations 1000-year celebration as the Kingdom of Norway. It was used for oceanographic research and was even ordered by the Storting to sail to the eastern part of Finmark to settle a conflict with Russia on the fishing boundaries between the two countries. This was a lingering dispute that had gone on for several hundred years and sometimes resulted in fights between Russian and Norwegian fishermen.

The *Hansteen* established where the border on land continued at sea and there were no disputes between the two countries for a hundred years. Only recently has a new conflict emerged on the "Gray Zone," which includes the whole Barents Sea, believed to contain vast oil reserves. Today, the dispute is no longer the fishermen's fight, but a tussle between bureaucrats and politicians.

Hansteen's depth soundings of the fishing banks that also established the type of sea bottom was important for the growing fisheries that had moved off the coast. The new fishing maps became a blessing for the fishing fleet of open, semi-decked and decked sailing smacks used when fishing on the banks in those days. G. O. Sars, one of the forefathers of modern oceanography, did research on the cod from the deck of the *Hansteen*. On one particular research voyage the *Hansteen* was sailed all the way to Scotland.

En Godvejrsdag paa Storeggen.

This xylograph from Skillings Magazine, 1870, depicts the *Hansteen* surveying the bank of Storegga on a sunny summer's day with fishing vessels all around. Sailing smacks was often used on the banks.

In 1899, *Hansteen* became a local fjord steamer for about a year but did not function well as a passenger ship. People had gotten used to quality passenger steamers, and newspapers printed complaints about this outdated survey vessel. This ship was too old and was built for an entirely different purpose. Then it was sold to Sandnessjøen in the north, where it was rebuilt and renamed *Haarek,* and converted to a ferry. From 1900 to 1948, she was in service on the fjords and islands of Helgeland. Old *Haarek* should have been scrapped but ended up at Florø, got a barrack on top, was renamed *Ivar Elias* and became a dining hall and sleep-over place for workers in the herring industry. By 1962, the herring fisheries had culminated. The barrack ship was towed home to Oslo and became a shelter for the homeless. In 1978, the shelter was closed. She was slated to be burned for pyrotechnical research when a private party saved her for restoration. At that time, the lower part of the ship was in the best shape — with doors, stairs, walls with corners, cabinets, and washstands in good condition. Bunks for navy sailors still had visible original numbers.

From 1979 to 1986, the hull and exterior ship was brought back to its old appearance. Veteranskipsreparasjoner A/S did the riveted ironwork. By 1993, the ship was completely restored to its original glory as a research ship and to an excellent and seaworthy condition.

Hansteen at Tromsø about 1890, manned by the Navy. Notice her furled white sails and the union flag. This was the ship we recreated by removing several versions of a local fjord steamer and hostel built on top, and then restoring the lower and complete part of the ship. Inserted is *S/S Haarek*, steaming along the coast of Helgeland before WW II. A complete weather deck had been added along with new superstructure on top of *Hansteen's* main deck that had been lowered fore and aft. Later a barrack was added on top.

Note for the English edition

Since the iron research and restoration was finished in 1993, *Hansteen* has been sitting in a harbor without professional care. It has not been covered up as needed during harsh winters. Hatches were closed and heat became trapped inside with rain and snow on deck. New decks have developed rot and leaks. The decks have been salted; salt has penetrated into the ship together with rain water. The hull's plating was wrongly measured and iron pieces in excellent condition were replaced with steel. Finally, the iron was sandblasted with steel grit, damaging twenty iron plates.

The ship is in need of a museum with qualified persons in charge. Authorities are trying to force this historic vessel into service for the tourist industry during a few summer months, and have replaced originality with today's off-the-shelf items from parker-screws to radar. Old *Hansteen* gave the answer herself to the idea of steaming with passengers a hundred years ago. In 1899, she proved absolutely unsuited for this type of service. The same thing is happening in the same town today — she is made to work again while her authenticity is being sacrificed. Taxpayers' money is spent to destroy this icon rather than to preserve it. More than 17 iron plates have been replaced so that the ship is able to meet improved requirement from the certification officials. (K. Petterson, personal communication, 2003).[5]

[5] Visit to Bredalsholmen, Kristiansand, during hull repairs on *Hansteen* in 2003, and was informed that 17 whole plates and some other partial repairs on iron plates had been done to the ship during the last year.

HISTORIC REVIEW

Of the huge number of ships made during the iron époque of the 19th Century, only a few relatively complete iron hulls in seaworthy condition still remain. Some former iron ships have been repaired and restored excessively and appear today with a steel hull. These ships are lost as genuine historical documents. Wrought iron ships with good and relatively original hulls are rare.

For commercial shipping, iron has not been used as a ship building material during the last hundred years. The old hammer ironworks were closed down when better quality rolled ship building steel was produced in great quantity by new methods. This new product captured the market, and steel became the shipyard's preferred building material for the production of ships. Modern roller steel products have been in use for ship building ever since.

The last existing ships of iron represent a treasure that should be saved and protected. Once the iron has been replaced or damaged, the value is lost and a piece of history would be lost forever.

HULL REPAIRS 1980–1986

The first surveying of the floating barrack *Ivar Elias* ex *Haarek* ex *Hansteen* was made in 1978 before any decision on saving and restoring this ship had been made. A surveyor from Det Norske Veritas measured the hull of this ship with an ultra-sound meter to establish the status of the hulk and evaluate if the ship was worthy of saving. After a few measures he could tell that there was plenty of iron left, and before the end of this preliminary investigation, the surveyor was very impressed with the readings he got. He repeatedly stated: "This old ship is amazing. The hull has plenty of thickness left. My readings far exceed the minimum thickness for this ship's small tonnage" (O. Joki, personal communication, 1978). This indicates that the ship was over-dimensioned or very strongly built. The controller approved the plating with acclamation.

This corresponded with what I knew from researching the building records and her history, which included reports that she ran aground at full speed several times on hidden rocks during her years of depth sounding in unknown waters outside the coastline of Norway. In two of these instances, she ran so high aground on a submerged rock that she was left dry during low tide. Both times she was pulled off undamaged and did not leak.

When the restoration of the *Hansteen* started, extensive new surveys showed that less than one out of every fifth plate had to be replaced. Most of those plates proved to be secondary or had secondary partial repairs. Those were repairs for recorded damages like stranding or collision, primarily from her 50 years as a local steamer. A few old plates had been repaired temporarily by welding during the later years. All repairs had been done with steel plates of less durable quality compared to the ship's original hammered iron. Steel had rusted away or been "sacrificed" like some rivets in her forward side which looked like they had functioned as anodes. None of the ship's original iron plates that had no earlier damage showed measurements that were less than what was required. Some readings we got on the ultra-sound meter with less than the required thickness proved to be a result of impurities within the plate. These hull plates were good and were kept intact in the ship.

Left: Last docking at Nyland's old stone dock. We docked the three historic vessels ourselves. In front was the *Hansteen*, followed by *Oscarsborg I* (1904) and *Styrbjørn* (1910). The two latter are both made of steel.

Right: Documentation of the restoring process included still photographs and film. Shipyard workers are discussing the next take. From left: Wilhelm Trygg, Per Overrein, Terje Edvardsen, Gunnar Pettersen and the photographer Thomas Diseth by the camera.

After the ship's hull repairs had been finished, the *Hansteen* was still to a large degree an original iron hull underneath her main deck. This was cleared with the Norwegian Maritime Directorates local surveyor that approved her hull to be in good and seaworthy condition all over. We then determined that no more plates were going to be touched as her fine hull makes this ship unique. The inside angles and her five bulkheads were inspected and the plating was tested by ultra-sound anew and approved by the same ship surveyor. He also did inside pressure tests with a fire hose and approved all her numerous 5/8 inch certified rivets that had to be replaced by a licensed riveter. The riveter was extremely proud when none of the many hundreds of rivets showed a leak. The hull was declared good and sound, and measured more than 9, 8 and 6 mm throughout, moving from the keel and up. She then left her dry-dock as a riveted ship all over. Boiler surveyors from Det Norske Veritas pressure tested the boiler with cold water and found no leaks.

At this time, we declared that with proper sailing, handling and maintenance, the *Hansteen* will last for a generation or more as a steaming historic vessel before it has to be hauled ashore and protected inside a structure. By then, her work in research and education will probably have earned her a boathouse. It was also put in contract form that the ship should be maintained and kept in order, and that her look and appearance should not be altered. The *Hansteen* should be kept the way as it was during her surveying period in the 19th Century. This was even documented into her deed before I gave her away.

During the restoration of the hull, samples from her damaged and replaced plates and frames were cut out, labeled and sent to the laboratory at Norsk Jernverk for analysis. We also sent samples of what we believed to be secondary plates, for the records, and to keep the material conclusive. These samples proved to be of a different quality, and indeed, were modern steel replacements as expected (Nyrud, 1982).

IRON RESEARCH DURING THE SHIP'S RESTORATION

We were very interested in acquiring knowledge of the ship's building material during the process of restoring her hull. The goal was to restore this ship in the best possible manner from a historic and scientific point of view, and only replace parts that were required to produce a good and seaworthy vessel. We also wanted to learn as much as possible about the ship and its history. Laboratory testing of the ship's iron would be a great help in finding out more about the *Hansteen*. The following is what we found with regard to the original building material of this wrought iron ship.

BLOOMFIELD IRONWORKS

According to the iron ordering books of Nyland, Bloomfield Iron Works was a principal deliverer of most of the iron used in the construction of hull # 11, to be named the *Hansteen*. In the mid-1800s, Bloomfield was among the most famous ironworks in England in the production of iron material for ship building. Bloomfield exported its ship-quality iron all over the world. The Bloomfield Works, even if located in Staffordshire, aimed to produce Best Yorkshire Iron. This was a well-known, top quality, brand. When *Hansteen* was being built in 1866, Bloomfield had four hammers and 57 puddle-ovens. The weekly production was about 500 tons, with a 12-hour workday. After 1880, production decreased rapidly and the Bloomfield Iron Works closed down and was finally dismantled and sold for scrap in 1904.

Only five works on the British Isles delivered Best Yorkshire iron and demanded a higher price. Low Moor Iron Works just outside Bradford in Yorkshire was one of those five. The shipyard's records show that Low Moor iron was also used in the *Hansteen*. This ironworks had its best years of production between 1860 and 1880. Low Moor employed more than 3,500 workers. Production exceeded 600,000 tons a year. It had the world's biggest blast furnace, which was 23 meter high with an internal diameter of six meters.

A. THE STAMPS

Stamps found on the ship's angle iron and bulkhead during the restoration period were identified and found to be consistent with the Nyland Værksted's record books for the ordering of iron from England for her building. Yard order records from the two English ironworks Bloomfield and Low Moor show that both ironworks contributed materials for steamship # 11.

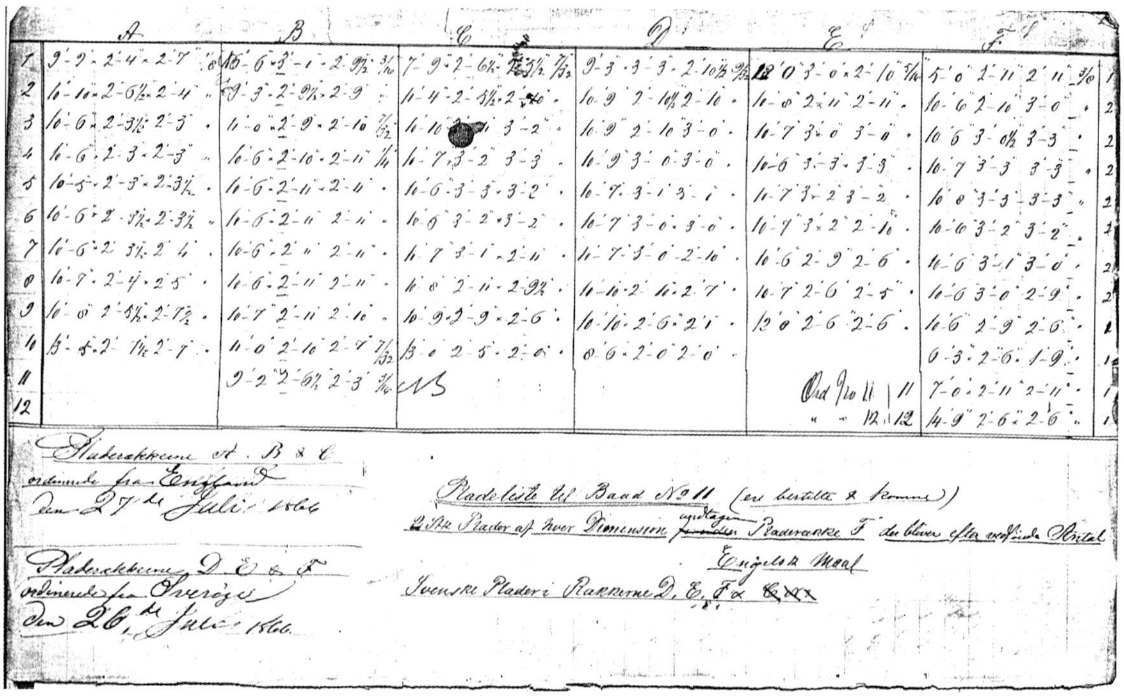

Printout from ordering-book A (Materialbog A), of Nylands Værksted, 1862-1877. Plates for the *Hansteen*.

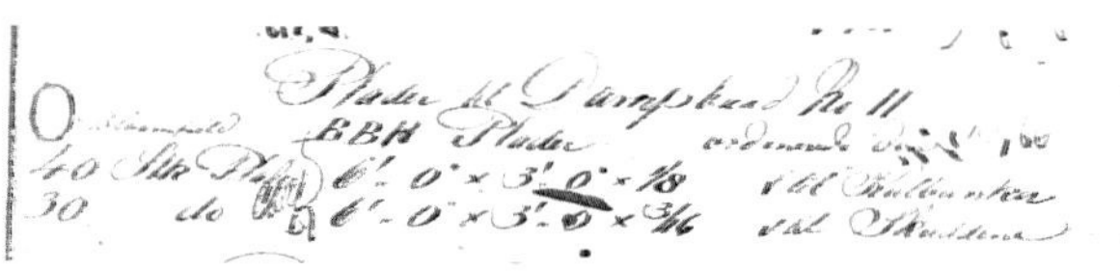

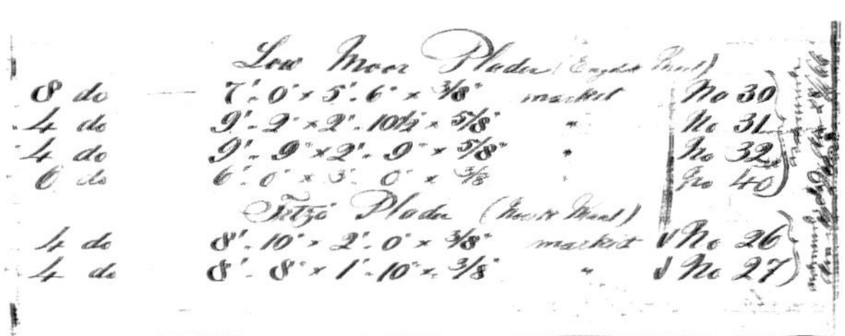

Producer's marks on one of the named plates (above right) are still visible aboard ship # 11 (see image on next page).
Right: Every piece of iron ordered and used in a ship was recorded with the ordering date, when it arrived, and where it was used. The samples shows iron from Sweden, Low Moor, Fritzøe and Bloomfield (BBH) Iron Works.

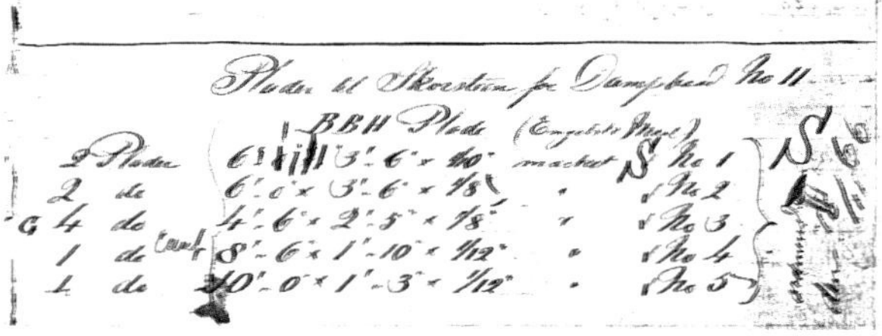

The BBH BLOOMFIELD Plate Stamp

During work in the engine room, I came across an imprint on the aft bulkhead's port side between angle 1 and 2, about one-third up. It read BBH with the word BLOOMFIELD underneath (see image). All the markings were capital letters indented in the plate. The height of each letter was 10 mm. The length of the name Bloomfield was 50 mm. The separating distance between the two lines was 6 mm. This was, indeed, the proof of originality we had been looking for, namely, iron plates delivered by Bloomfield Iron Works.

BBH stands for Bloomfield's Bradley, Barrow & Hall, of which Joseph Hall probably was the most famous. He was an inventor and made a name for himself in the history of iron production. The imprint was a significant identification mark for the top quality iron from this ironworks.

The BBH.BEST Stamp on the Angle Iron

In the same general area as mentioned above, close to the upper end of angle iron number two, phasing in and on the inside of the step of the angle, I found a convex impression reading BBH.BEST facing up. These were also in capital letters but slightly smaller than the ones on the plate (see image). Two extra B's can be seen in the front of the name. It is hard to say if this is a misprint, or if they indicate that this angle is of BB or Best Best quality. We know that the Bloomfield Iron Works used the different quality grading Best, BestBest and BestBestBest for their iron products, where triple B designates the most superior and expensive quality hammered iron produced by the company.

1866 BBH BEST mark found on the angle iron.

1866 BBH BLOOMFIELD mark found on the plate.

According to the ordering books from Nyland, Bloomfield did deliver Best Best Bloomfield Ship Quality Iron to Christiania for the construction of yard # 11, to be used for the frames and beams, and for support of the five bulkheads. Plates for bulkheads, coal boxes and hull plating for three of the six overlapping streaks of hull plates in the *Hansteen* were all from Bloomfield.

I was told that new iron plates and profiles produced at Bloomfield had to sit outside under a roof in their lot for up to two years without developing rust on the surface before it was sold and exported as best ship quality building material (Larsstuvold, 1982). One hundred and fifteen years later, we have strong evidence that this could be a true statement. When we broke loose the concrete in the bottom of the engine room of the *Hansteen* and exposed the inside hull plates underneath, we witnessed, for the first time since her building, beautiful gray iron plates with no rust. This iron had never had rust on its surface and was in the same state as when laid down and cemented in 1866!

It is unbelievable that the iron plates could be as good as new in a 115-year-old ship that had sailed the North Atlantic Ocean back and forth between Sandnessjøen and Træna, an island off the Atlantic coast of Norway, for almost 50 years after her 30 years in offshore research was over.

B. LABORATORY TESTS ON SUBMITTED SAMPLES

During the restoration, we replaced entirely or partially 26 plates in the hull of the ship below the main deck. Early on during this process, we realized that many of the plates in question seemed slightly different in texture from other good and undamaged plates in the ship. By studying old records and examining repairs to the inside framing we could determine which plates had been repaired or replaced. Oddly enough, the secondary and newer plates, supposed to be of a better quality material, were in poor condition compared to the good ones we presumed to belong to the ship's original iron hull as they showed no sign of repairs. Riveted repairs are easily spotted in a riveted ship. Repairs by welding was almost non-existent as the ship was taken out of service while riveting was still the only certified way of repairs in Norway.

Between 1980 and 1982, we cut out and labeled 13 samples from the frames and the plates we took out in different parts of the ship on both sides. We did the same in 1982-83 for 14 new samples taken from the rest of the replaced material. All these 27 samples were shipped to the laboratory of the A/S Norsk Jernverk at Mo on two occasions, with a request for an analysis of the submitted material. It resulted in two test reports. The first one by laboratory engineer Roald Nyrud and the second one by laboratory engineer Svein Johnsen. The first one is dated March 10, 1982 (16 pages) and the other is dated July 3, 1984 (9 pages). Both reports are written in Norwegian.

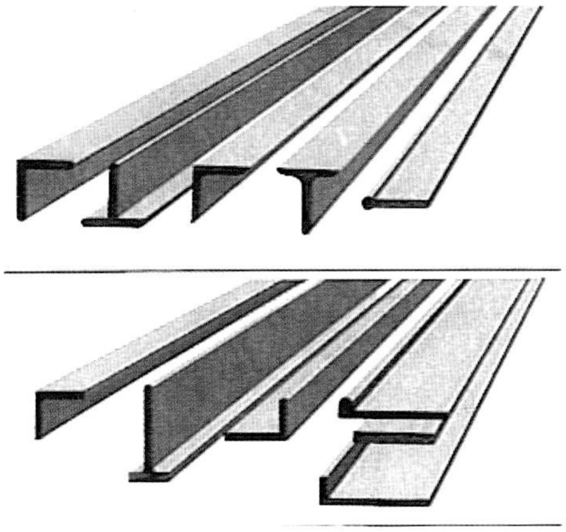

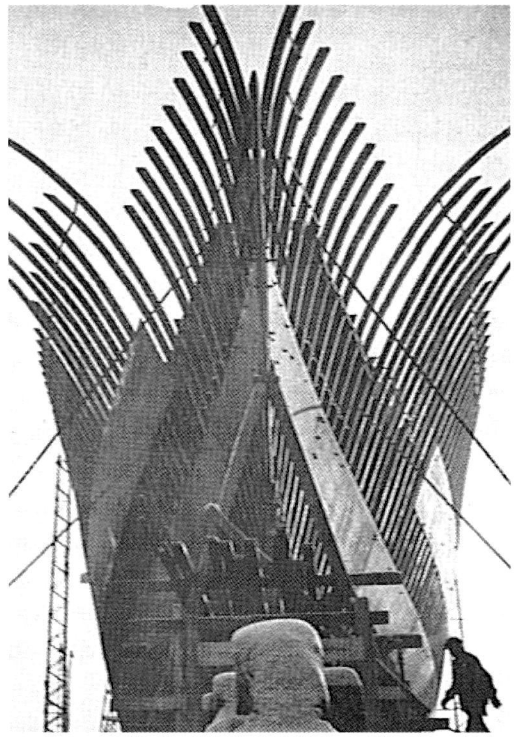

Above: iron profiles from the time *Hansteen* was built. The lower are of steel used in ship building today.
Right: a ship under construction at Nyland. This illustrates how a riveted ship is put together. Iron and steel hulls were assembled in almost the same manner as large ships made of wood.

Roald Nyrud's report from 1982 starts with a definition of the job, to test the samples from the hull of the *Hansteen*. The submitted samples would be analyzed with respect to the steel and iron manufacturing processes known when the ship was constructed. Tests were conducted to see if the samples comply with them or not. The laboratory tests aimed to find out if the samples are from the original hull or if they are replacements done at a later point in time.

We submitted 13 samples for the first laboratory analysis. They came from five plates removed on the starboard side, (marked sb) and five from the port side (bb). We also added three samples of the frames (sp) from the aft part of the ship where they were in a bad state of decay. Two came from starboard and the last from port side of the vessel.

Nyruds's conclusion after several tests is that the three angle irons are all from the original construction, probably produced in England by the Puddling method. Only four of the samples from the plates belong to the original hull. They are produced in Sweden by the Lancashire method. When we received the test report, we could confirm that they all came from the three lower streaks, said to be of iron from Sweden. The rest of the samples come from plates that must have been placed in the hull later by repairs. Variations in the analysis suggest that they were attached at different times in the ships history. Low content of nitrogen shows that none of the plates are of Bessemer or Thomas steel. They have to be either Siemens-Martin or electro-steel.

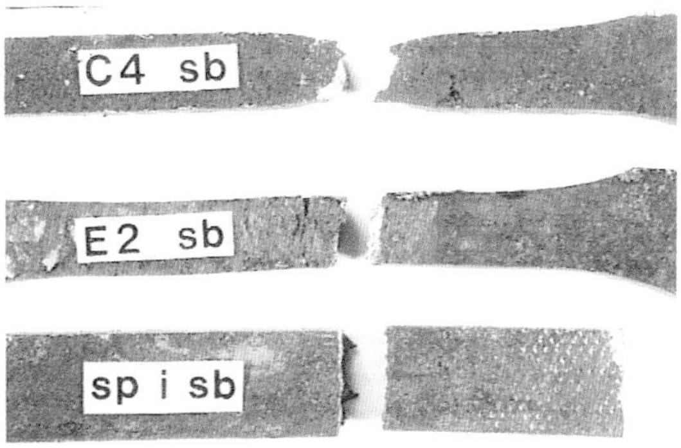

Fig. 2 Strekkprøver

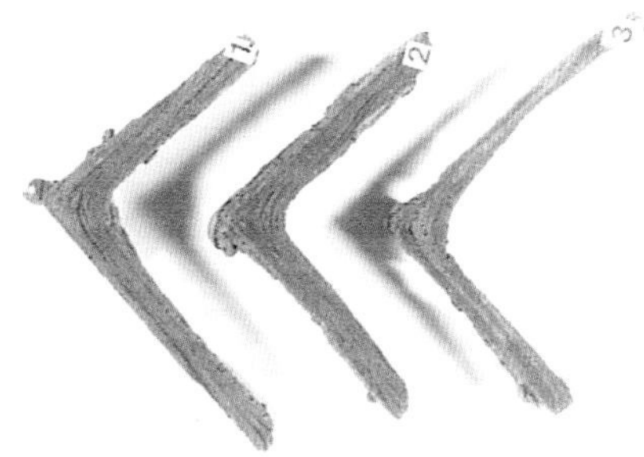
Fig. 5

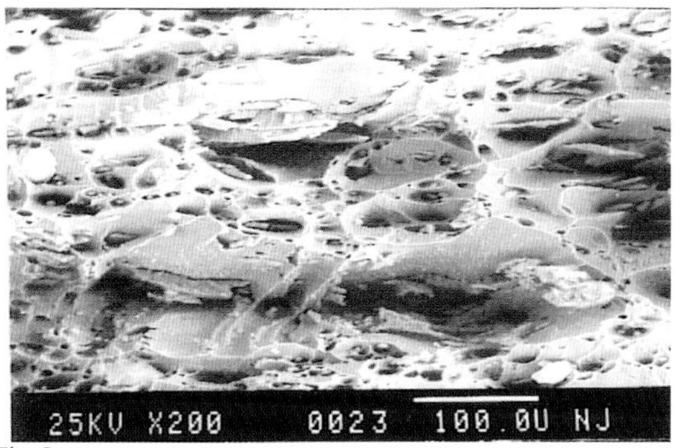

Fig. 3.

Fig. 6. 160x

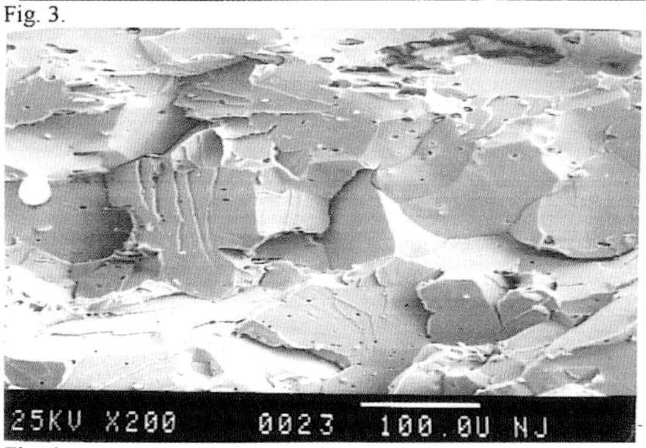

Fig. 4

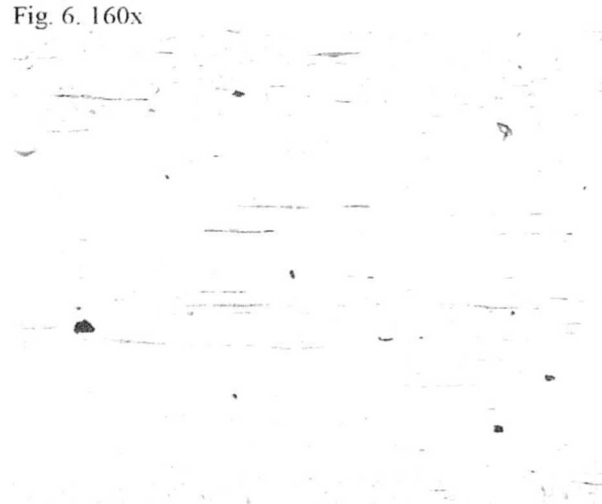
Fig. 7. 80x

Four different tests were done at the laboratory. A piece of modern ship-steel (A-36) produced by Norsk Jernverk was added to the sample as a reference. The tests were:

a) Chemical analysis;
b) Tensile test;
c) Deep etching (of the frame samples); and
d) Metallographic analysis.

During the chemical test, the samples were checked for percentage of carbon (C), Si, Mn, P, S, N, Al. The table shows that some of the samples have an extreme low content of carbon. The three frame samples have a very high percentage of phosphorous.

The tensile test showed a great range in durability (elongation) (see images above) from six to 22 percent, while for the frame samples there were impossible to record any result due to the content of slag (Figs. 3 and 4). They were then deep-etched and examined under a microscope (Fig. 5).

Fig. 8 50x

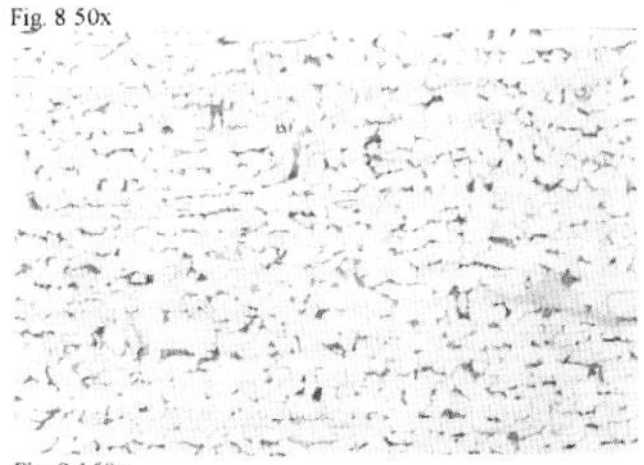

Fig. 9 160x

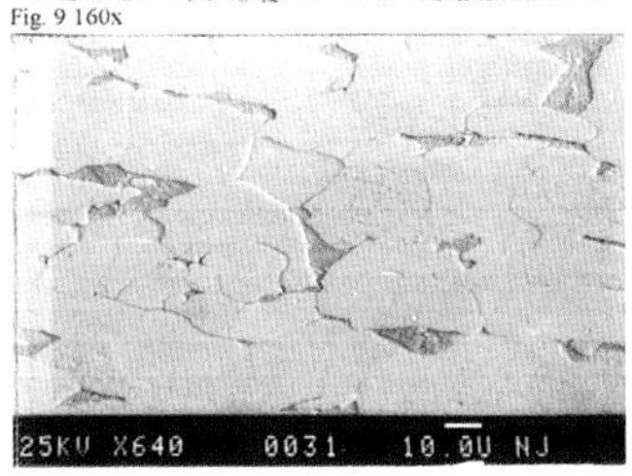

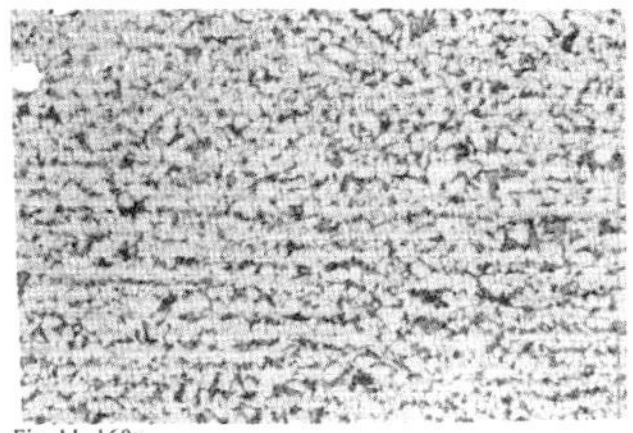

Fig. 11. 160x

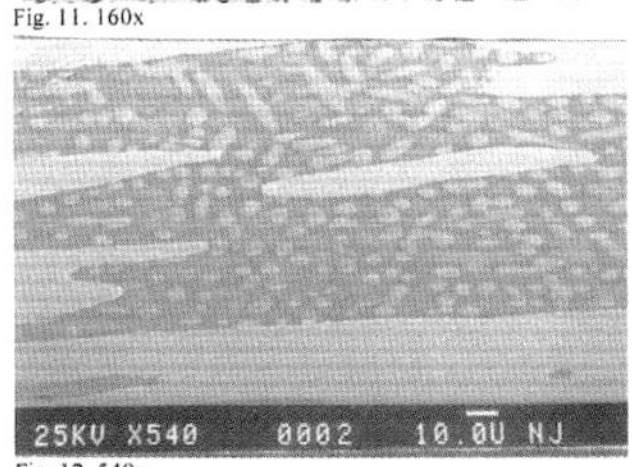

Fig. 12. 540x

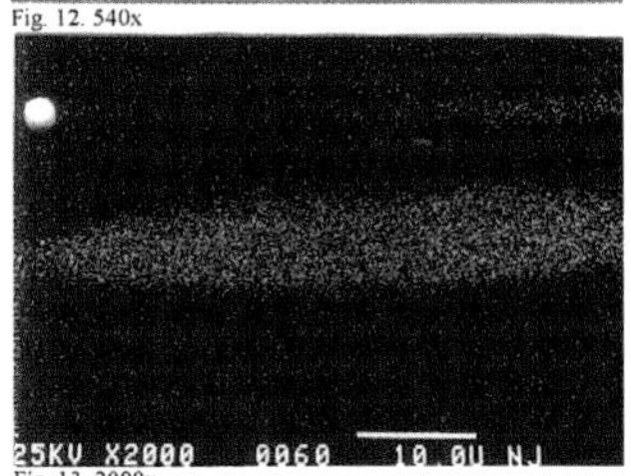

Fig. 13. 2000x

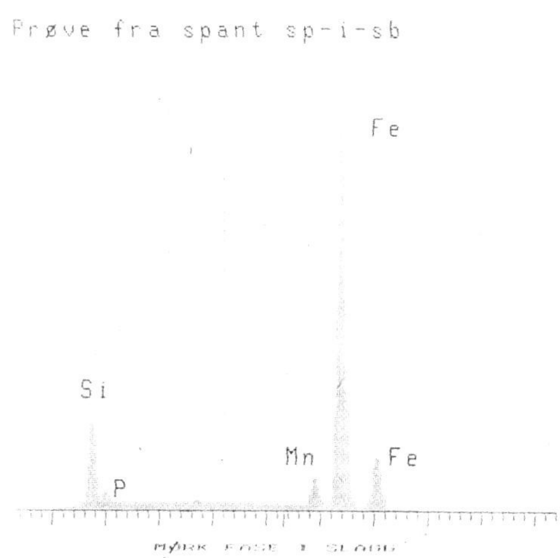

Fig. 14

The laboratory made metallographic sections from cross-sections of all the 13 samples. The three frame samples and four of the plate samples had a considerable amount of inclusions, probably slag from the production (Figs. 6, 7, and 8). One of these (sp-i-sb) was examined under the Scanning Electron Microscope. Figure 12 shows one of the two phases. A qualitative analysis of these two phases (the light one and the dark one) shows Fe and Fe-Mn, P, Si in the dark phase. Most of the phosphorus found in the chemical analysis is probably from the inclusions. This is typical for hammer iron. All the other samples show lesser amounts of inclusions and of a different type, basically manganese-sulphide. The x-ray mapping of one of these is shown in Fig. 14 (O. Handaa, personal communication, 2004).

Comments by Nyrud indicate that the samples are of different qualities. The three frames and four plates are different from the others. The rest are more normal and shows a lesser content of trace elements, even if the analysis shows quite a bit of variation. Clearly, they are produced with more modern processes. The low degree of phosphor and sulfur in the four plate samples is due to cleaner iron ore and the use of charcoal during the production. This indicates that it is made in Sweden with its high quality ore using the Lancashire method. None of the other plates could have been produced in England by a Bessemer converter. No deliverer was given for the three frame samples, but the very high degree of phosphor indicates that phosphor-rich iron ore has been used. It is therefore assumed that this iron is produced in England by the Puddling method. Only these three samples plus the four plate samples come from the original produced hull.

We shipped another 14 samples we had been gathering during the latter part of *Hansteen's* restoration from the spring of 1982 until the fall of 1983, to Norsk Jernverk. This time they were analyzed by Svein Johnsen.

Johnsen's report in 1984 referred to earlier reports, tables and figures. In a short commentary, he concluded that the (14) samples of submitted iron belong to two main categories. Sample 3, 4, and 5 are original hammered iron, produced in Sweden by the Lancashire method. (Those plate samples all came from the beaten-up bottom of the hull). Johnsen states that they are of the same quality as the ones tested two years before. All the other (11) samples are of modern steel. They belong to modified sections of the ship (including a part of the keel shoe from underneath the engine room, ice half-staff at the waterline forward on the starboard side, and a knee aft on the port side).

Johnsen's, as well as Nyrud's, laboratory results were consistent with the records from Nyland, of which the laboratory had no knowledge. Their reports confirmed our knowledge on the repair done on this vessel. Those would be the replaced steel plates, along with new frames representing repairs from recorded collisions and other structural repairs. The vast amount of available written material, from yard requisitions, written records and casualty reports, to books and magazines, all supported our theory that we worked with during restoration period. It all points to one conclusion: that the untouched plates left in the hull of the *Hansteen* when the restoration of the ship was finished in 1993 consist of original hammer iron from the construction in 1866.

Hansteen is a unique example of very early Norwegian iron ship building, which represents many new opportunities for research and scientific studies to further extend knowledge of this first period of iron ship construction and design. It is a field of international interest. Wrought iron structures from the classical era of the 1860s are rare, ocean going and coastal vessels still in service are extraordinary items. *Hansteen* is a unique sail and steamship and could be the oldest ocean going steamer built of iron that is still under steam.

Stepping back in time. The roof of the barrack (*Ivar Elias*) was lifted down and became shelter while burning away the steel weather deck (*Haarek*). All this and 40 tons of gravel in the bottom made the black part (*Hansteen's* iron) slowly emerge from the water as we lightened the hull.

Under sail in the Pacific. *Star of India* (1863) is a genuine and unique iron ship. She is the oldest major vessel in the world still sailing on her original hull. Photo: The Maritime Museum of San Diego.

STAMP MARKS FOUND ON BOARD IRON BARK *STAR OF INDIA*

The ordering lists from the Gibson, McDonald & Arnold Shipyard at Ramsey for the iron used to build the ship *Euterpe*, now bark *Star of India*, in 1863 are all lost together with most of her other building records. My investigation shows that the ship's frames or ribs, beams and angle iron are all marked with LW & WALKER and CONSETT, telling us that this ship was made from these particular brands of wrought iron. The findings are in part consistent with the official survey done after the ship had been completed from the yard and was ready for her sailing days in 1864. This report reveals that the *Star of India* is an authentic iron vessel with the original wrought iron hull from her construction still intact as far as I could tell (Engvig, 2000).

INTRODUCTION

The main objective of my investigation, which took place on the 10th and 11th of April 2000, was to find marks made by the iron manufacturing company for identification, quality and originality reference. Identification of iron markings gives a platform for research into different ironworks, production of ship quality iron, how these products were made, shipped, used, and how well they lasted.

Furthermore, such markings serve to identify the ship itself. They provide proof of originality, indicate the quality of her construction, and offer us knowledge about the ship's service and maintenance over its extremely long lifetime in a hostile environment that deteriorates iron. Most ships last less than thirty years. In view of that, what makes a few ships survive more than one hundred and thirty years?

My goal was to try to find impressions similar to the ones I found when doing restoration and research work aboard the Norwegian wrought iron ship the *Hansteen*, also a seaworthy vessel, built at Nyland, Oslo in 1866 of iron from

England and Scandinavia. This could contribute to a comparative analysis of ship quality iron from ironworks in Great Britain, Norway and Sweden.

Scandinavia and Great Britain used different quality iron ore as well as production material and methods in the wrought iron production processes. These countries also have an old iron and ship building tradition with the use of wood and iron in combination that dates back to the Viking Age and before. They also produced and used industrial iron for an iron ship building industry early on in the 19th Century when this type of ship construction started. Before the end of the same century, iron ship building was abandoned as new types of steel had captured the world market. Ships have been built of steel ever since. Of the thousands of iron ships built between 1840 and 1890, only a few seaworthy and original units are left.

HISTORIC REVIEW

The three-mast bark, *Star of India,* is the starter and the star item in the Maritime Museum of San Diego's collection. The ship is moored alongside the pier on Harbor Drive in downtown San Diego. The *Star of India* belongs to the classic period of iron ship building and has survived the ravages of time very well. As this report will show, her iron hull is practically untouched since she was constructed in 1863. That makes this ship remarkable. Other existing iron ships have seen damages, repaired and has been restored excessively, many from a hulk, with the use of modern steel. This has been done to meet today's demand for safety and aesthetics, leaving only few vessels with their original iron hull intact. The *Star of India* may well be one of the most genuine iron ships in existence.

The *Star of India* is the former full-rigged ship *Euterpe* of 1,197 tons and 205 feet, built at Ramsey, Isle of Man, UK in 1863 for general trade on all oceans. She was first employed in the India trade hauling general cargo to India returning to Great Britain with jute and other colonial products. After almost eight years in this trade, she was sold in 1871 to Shaw, Savill & Co for the emigrant trade on New Zealand and was refitted to carry more passengers. On her first run in 1873, she reached Dunedin after 117 days out. Spending almost four months at sea was considered a long time in those days. The *Euterpe* served well as an emigrant ship, but was one of the slowest of all the fine packets Shaw Savill had in this trade during the 1870s.

Specially made iron ships like *Crusader* and *Zealandia* often made the trip from England to New Zealand in 85 to 95 days, while the *Euterpe* would take closer to 110 days, or usually more (Brett 1924). But she was heavy-built and not a wet ship rigged with skysail like the *Zealandia,* a speedy packet with her deck regularly awash. In 33 years, she did 32 round trips from London — 30 of them lasting shorter than 110 days. Built the same year as the *Cutty Sark,* the *Zealandia* is the unknown legend of this trade. Even after being sold to Sweden in 1902, she continued sailing to New Zealand, carrying lumber. She was 42 years old when she stranded on Nova Scotia.

The postcard below from Le Havre, dated 1905, shows the speedy iron ship *Zealandia* arriving from New Zealand. Her long white poop deck tells us that she was made to carry passengers.

The captain and crew of the *Zealandia* when the ship is 36 years old and after she became Swedish. This photo was made at Wellington in 1905. Young Thormod Engvig is on the gunwale to the left. His first trip around the world was on board this ship.

The *Euterpe*, as slow as cold molasses but dearly loved by everybody (Bowen, 1939, p: 29), stayed out of trouble. Only minor incidents happened to the ship during her 27 years with Shaw Savill. Immigrants seemed to have no objection traveling on her. Landlubbers, afraid of the sea and liable to seasickness, often preferred a slow and sturdy ship compared to a wild race down the Atlantic. Making late arrivals, the *Euterpe* sometimes had problems finding return cargo as it would have been picked up by other faster vessels. She even had to go to the U.S. West Coast a couple of times to find return cargo. Eventually, *Euterpe* was known to the Americans, when the British owners wanted to sell her. Bound for Hawaii, outside U.S. shipping jurisdiction at that time, an Hawaiian-born American living in California bought her. He wanted her in the Hawaiian register and experienced some brushes with the authorities all the way to Capitol Hill. This was 1898 when the Hawaiian Islands were annexed by America. Some fine British ships made it into the U.S. registry in this way. From Hawaii, the ship sailed for British Columbia after discharging the coal and was dry-docked and inspected. By fall, the crew was paid off by the British Consulate and the last British captain left his ship (MacMullen, 1961).

She picked up lumber for Australia and this was the start of her lumber carrying days. She was to be used in the lumber trade from the Pacific North West to Australia together with other lumber ships from San Francisco. In 1900, after more legal battles, the short-lived Hawaiian-American ship finally became full-fledged American. Her time as a lumber, coal and sugar carrier in the Pacific was brief. In San Francisco, 1901, she was sold again, this time to the Alaska Packers, her last professional owners.

Euterpe at Port Chalmer. A photo of a whole series, taken after Shaw, Savill merged with Albion 1882, as the new name appears on the warehouse roof. This full rigged ship looks great, and a lot differently from her later years as an Alaska Packer. Courtesy Maritime Museum of San Diego.

The *Euterpe* sailed with the Alaska Packers of San Francisco until 1923. At this time she had 60 years of service on the high seas. In the spring of 1902, she left San Francisco on her first of many seasonal voyages to Nushagak River in Alaska for the salmon season during summer, down to her mark with stores and cannery supplies along with about 200 fishermen and cannery workers. It was a wild five-month affair with a mixture of cannery people, salmon fishers, masters and sailors, constructers, factory maintenance men, carpenters and engineers, all employed by the Alaska Packers.

When Shaw Savill and other big shipping companies in Great Britain and Ireland converted to steam and sold out their fleet of sailing vessels around the turn of the century, a considerable portion of them went to Norway. Many of their sailors discovered better pay and better food in America, jumped ship in San Francisco and signed on with the Alaska Packers. Even floating ice up in Alaska was something the Scandinavians knew from before.

While the *Euterpe,* now re-rigged as a bark, was in Alaska, the company was consolidating its tonnage by switching from wooden Down Easters to iron windjammers just like Norway had been doing. *Euterpe* and a few other ships had made the way into the American register for foreign built ships. By 1906 Alaska Packers had eight of them. Among them are *Star of France, Star of Russia, Star of Italy* and *Star of Bengal*. Someone came up with the great idea to rename the rest of the fleet with a star and a country. This was easier said than done, and bureaucracy and politics all the way to Capitol Hill had to be involved. Finally a bill was passed which President Theodore Roosevelt signed into law. This was the second time the *Euterpe* had been before the Congress.

Her prime award was the new name *Star of India*, a name she has carried ever since. It was painted on after she returned to Alameda for her long winter season laid up together with other new "stars" of the Alaska Packers. She continued sailing to Alaska for salmon until 1920 when she had to be towed all the way due to a lack of experienced sailors. 1923 was her last season, after which she was moored over at Alameda awaiting her destiny with the other "stars."

A late picture of the *Star of India* making sail as a commercial ship leaving Alaska in 1921. What a mess she was! She was 58 years old and carried some strange sails, her yard was tilted, and her crew was struggling to get the sails right. The captain had a hard time finding experienced sailors. Many were steamboat men who were not acquainted with a square rigged ship. Photo Otto Weidemann/San Diego Maritime Museum.

Her sailing days were over, at least as a commercial trader of the seven seas. The bigger and newer stars of steel were sold to Japan and melted down perhaps to become raw materials for Imperial Navy war ships. The *Star of India* was the old one, made of iron, and priced accordingly. When a group in San Diego who were interested in maritime history wanted a tall ship for a maritime museum, enough money was raised to by the least expensive "star." They even got a fine Viking ship as bonus but forgot to build a boat house like the Vikings did. The *Star of India* ended up in San Diego as a museum ship in 1927.

She has been there ever since and has had her ups and downs over the years. A favorable climate and a few persistent people have always managed to pull her trough crises and periods of neglect. Today she is one of the really well-kept museum ships of the world. She is in good care and even seaworthy. She could have been lost to the world when almost no one saw any value in her other than her scrap price. A few brave men in San Diego changed her fate.

THE SEARCH FOR IRON MARKS

The goal of my survey of the ship focused on:

1) An inspection of all her frames looking for marks or impressions.
2) The same survey for all her beams in her two original decks.
3) Her inside plantings, particularly the two bulkheads that could be inspected on both sides. (I assumed that impression on plantings was done on "the right side" which usually would face out or forward).
4) Outside hull plating. This would be demanding as most of the exterior hull is inaccessible for inspection (Former experience also tells me that the possibility of finding marks on the outside plates is rather slim).

STAMPS ON FRAMES

I started my work in the lower part of the vessel, as frames there tend to have fewer coats of paint on them. The frames in the ship's hold on its starboard side below the tweendeck looked promising. On Frame 25 aft of the forward bulkhead, I registered fragments of convex letters just underneath the deck, some illegible, others obscured by rivets. I found these to be repeated lower on the same frame. I discovered the same pattern on several frames aft of this first one, but was not able to make out a word from the letters. "Between-frames" or ribs that are common in the bow section of many ships continue aft on this vessel. They also carried the same markings. Not all of the frames were open for inspection. On the accessible ones, markings were placed on the inside of the angle iron (the frame) on its step, or angle; 90 degrees on the ship's hull. For strengthening of the frames, the angles have another contra-angle riveted on to it. The letters all appeared approximately in the center of the angle. All angles or ribs were marked with the same letters, about 20 mm in height and protruding approximately 1 mm. The interval before letters or stamps were repeated is about one meter. I recorded the findings.

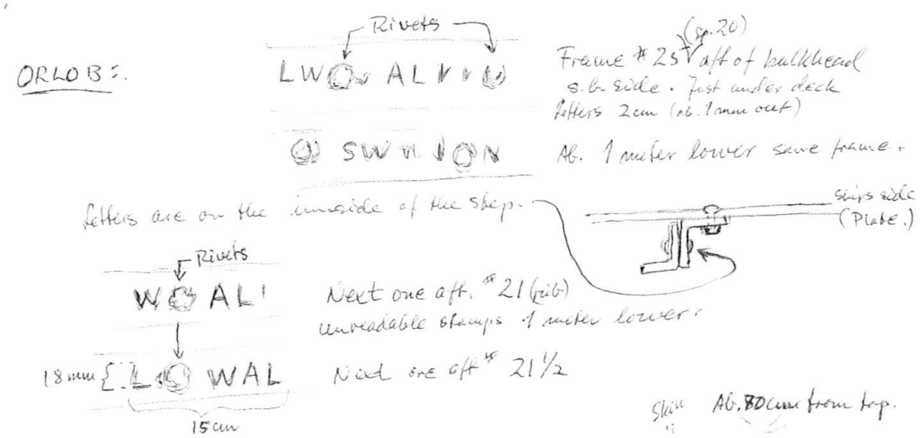

Sample from the first page of the recording of marks on the ship's frames.

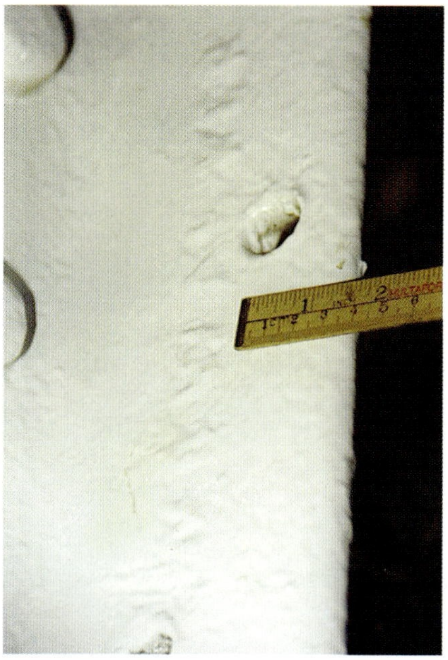

Forward port side of the lower hold with the bulkhead at right. The frames in this part of the ship showed many marks like the one to the left, they all have hardly visible manufacturing stamps from the Walker Iron Works, by Newcastle-on-Tyne, reading LW & WALKER.

After my survey of the orlob, I went on to investigate the rest of the ship's frames below deck, expecting the markings to continue up along the angel. Indeed, they did. I found the same stamps in the shop all the way forward. In the hold between the bulkhead and the foremast, I was able to follow the repetition of marks about every meter, going down along the frame to the cemented bottom of the ship. I found particularly well-preserved marks in the forward part of the ship on both sides. Display walls and locked cabins obscured much of the rest of the sides. I was, however, able to find the same markings on Frame 64 all the way aft in the ship. This made it immediately clear to me that I was identifying the ship's iron marks from 1863.

Some of the stamps in the hold and on the tweendeck were in excellent condition and not too obscured by rivets, or work marks from the ship's building, or later disturbance due to use, repairs or maintenance. The letters L, W, S, A, L, and at one point WAL were recorded by drawing.

By comparing partial stamps, I was able to combine the marks into a word that made sense. The registered S was obviously the symbol "&," thus leaving me with the manufacturers stamp LW & WALKER. (Could it be that two Walker brothers had owned an iron producing company and put their name on the angles to mark the brand or product for later identification?) I found additional markings throughout the ship. They all seemed to be of the same brand. Nothing I found indicated there was another name or another brand present. The stamps were so numerous and were of the same type. It made me come to the conclusion that there was only one producer for her original frames, whoever that might be. I finished my inspection with a satisfaction that I had found what I came for, and a notion that a lot of research over in England was about to start.

STAMPS IN BEAMS FOR TWEENDECK

I went on to investigate the beams, a sturdy flat iron with a bulb, and once again started in the hold below the tweendeck. On the first beam I found a clear stamp with the name CONSETT on the forward angle iron of the two angle irons riveted together on each sides of the beam itself to form the upper "T" for the deck to rest on and be fastened to. In 1863, due to the production processes for hammered wrought iron profiles, to roll or to shape a "T" like the ones in this ship on a bulb-iron would have been extra difficult. The procedure was to secure two angle irons with rivets going through the top back of the beam itself. The stamp marks were not on the beam itself, but on both the angle irons on the vertical part of the angle riveted on to the beam.

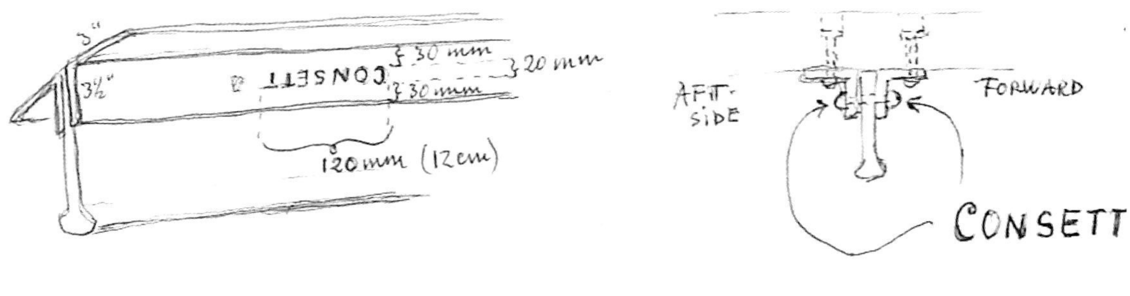

Above: Sketches of marks on deck beams beneath the tweendeck.
Below: Beams under the main deck and to the right the marks right above the ship's identification tonnage imprints, always found on the beam of the main hatch in any cargo ship.

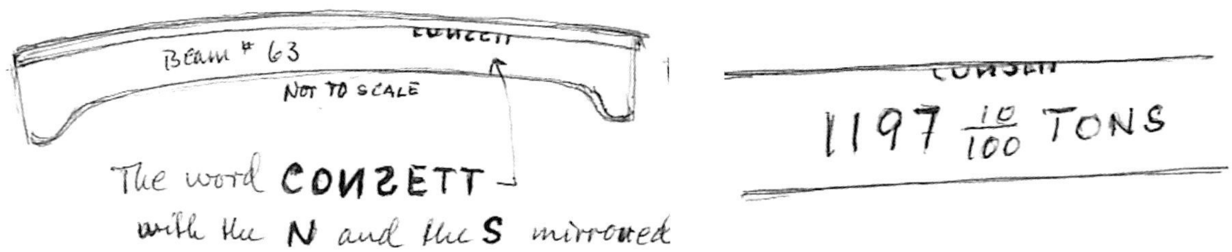

The letters in the word CONSETT were placed in the middle of the angle with the letters upside down or to be read from above, leaving about 25 mm of space between the letters and the inner corner, and between the letters and the outer rim, respectively. The letters in the word itself were 18 mm tall, thus making the vertical angle about 80 mm inside. The whole name was 120 mm long. These stamp marks were repeated seven times across the beam of the ship (35 feet). They were approximately 1,530 mm (1.53 meter) apart.

This stamping was found on all the angles of the deck beams that I was able to inspect in the lower part of the ship. Many of the stamps were in excellent condition and had not been covered with many layers of paint over the 137 years since the ship was built. The immediate conclusion is that the ship's tweendeck is original, and has never been altered or repaired with later quality iron or steel.

The first beam I looked at under the tweendeck (beam #23 aft of the bulk- head) had this fine CONSETT mark and I photographed the shown image (Enlarged above).

BEAMS FOR MAINDECK

When I started investigating the deck beams on the main deck, I did not find any marks in the angles that were riveted to the beams in the same way as on the deck below, and this puzzled me. Also, the shape of the beams themselves was amazing. Someone had produced the beam and its knee, a triangular iron plate at each end, in one piece, instead of using the common junction of the frames to the deck beams with a simple knee plate (similar to ships of wood). This is the common junction in riveted ships. It made me believe that all the beams (which are actually a bulb iron with two angle irons to form the "T" on its back) were each produced based on individual specifications to fit the beam of the ship as she narrows towards the bow and stern. Was this part of the ship's construction at the ironworks? Three years later on the *Hansteen* in Norway, this work was all done at the yard from standard size iron plates (knees) and angles.

I started my work on the deck beams on the main deck all the way aft. No stamps were found. A few irregularities that appeared to be drip marks from paintings seemed to be too regular in intervals to be just paint. After close investigation I found it to be the lower one-third of a name sticking out from underneath the rim of the riveted on angle irons, to make the "T" on the deck beam. It was covered with several thick layers of paint.

Moving forward, it became clear that underneath the main deck, the beam itself had been stamped with the CONSETT markings, telling me that some company named Consett made these special beams. The person administrating the marling did not know that they would be obscured by the later added angles or he would have lowered them, probably this was the distance from the upper edge stamp moldings was set in all their production tools. The beams for the main deck were marked in this manner all along the ship. I found stamps on almost every beam I inspected from all the way aft to all the way forward.

I counted six repetitions of the stamps in some beams, at about 1,600 mm (1.6 meter) intervals. The word CONSETT measured about 140 mm. Often, about 10 mm of the lower part of the word would show, while some places aft only about one-third of the letters height would show. In a few places, I could find the greater part of the name exposed for example on Beam 63 (see sketch page 146).

Unlike the lower deck where I found names stamped upside down, stamps under the main deck would appear the right way. One peculiarity puzzled me. The name would usually be spelled CONSETT, but I also found the name written with the "N" turned upside down and also with the "S" inversed, thus resembling a "Z." I even found one beam that had both the "N" and the "S" the wrong way, but the other letters all being correct. Since the letters are convex, or higher than the iron, the production company must have had the letters set in their fabrication tools. Maybe some of their able craftsmen providing the letters were illiterate? They were probably not made in that way on purpose, but by mistake as they had to be mirrored to appear right.

Right above the ship's identification marks in the middle of the beam on the aft rim of the main hatch is another stamp of CONSETT. This impression is partially covered by the angle iron as most on the main deck beams appears. Some of the CONSETT stamps on the beams were in excellent condition and not too much obscured by work marks, rivets or later disturbance due to use and maintenance.

My thorough investigation into the iron on the *Star of India* and the analysis done on the sail steamship *Hansteen*, built in Norway three years later, told me that this was British iron, made by the ironworks that delivered the material to the constructing shipyard. Stamps in the *Hansteen's* iron show it was delivered from the Bloomfield in Staffordshire and was marked accordingly as BBH BLOOMFIELD and BBH BEST.

The conclusion of my investigation of the deck beams for the main deck in this ship is that all the beams were laid down during the ship's construction, and that they are still in their original location. The *Star of India* is surprisingly complete and original.

PLATES AND ANGLES ON BULKHEADS

This part of my investigation was the most complicated. The large collision bulkhead forward, aft of the chain locker that closes off the hold forward, looked good from the hold, but I was not able to register any impression on this surface. The forward side below the tweendeck was rusty and its condition was not as good as the rest of the ship, which appeared extremely well maintained. No marks were seen. The upper area was partly obscured by obstacles. The same was true for the bulkhead aft. This bulkhead, however, appeared in very good shape, and I was particularly impressed with the aft peak. The paint looked very old and the whole room was nice and dry and in good condition. No marks were found on any of the plates or bulkhead angles.

I also did an investigation of the inside hull on the tweendeck where I could get to the ship's side. Again, the hull looked good, with lots of paint and no signs of impressions or factory marks. Below the tweendeck, only a few spots could be inspected and nothing was found.

I finally did an inspection of the small portion of the outside hull that was against the slop tank spacer of the pier, from the deck-line and down. The ship's side had the usual wrought iron pitting that would probably have done away with any impressions years ago. My search for manufacturer's marks on the plates yielded no positive result. But the dot marks for painting of the load line symbols were all there. They would probably have been redone several times during the ship's years in active service.

My personal experience is that it is much more difficult to find marks on the plates than on angles and beams. First of all, it might be that each whole plate was given only one or a few quality stamps by the manufacturer. This imprint might already be lost in the process of building the vessel. If not, it would be lost due to later corrosion, or wear and tear of the hull's surface, leaving us with laboratory research as the only option to determine the production methods and thus the time and the place where the iron was manufactured.

After finishing my inspection on the ship, I looked at a picture of an impression in the hull's outside plating at the San Diego Maritime Museum Library. It was found during dry-docking and complete cleaning of the floating body of the ship in 1976. The location was believed to be forward on the starboard side below the waterline. This picture shows an impression quite similar to the BLOOMFIELD plate impression of the *Hansteen*. The one on the *Star of India* reads: BRUNSWICK BEST. This imprint is probably a plate quality mark.

This BRUNSWICK BEST plate stamp found on the ship's bottom during dry docking in 1976 is consistent with the information written by Lloyd's surveyor when the ship was ready for sea in January of 1864. Courtesy Maritime Museum of San Diego.

The library also holds a copy of the report on the ship's final survey, dated January 5, 1864, after the ship's completion. It gave a recommendation to Lloyds that the new vessel could obtain the highest class with Lloyds of London for insurance purposes. The surveyor stated: "She far exceeds the requirement of the Rule for the tonnage." The agency's committee approved her for the class one week later when all dues were paid. It was also reported that good paint and Portland cement in the floor of the bilge protected her iron. The report gave all the dimensions of her frames, angles, beams, hull plates and bulkheads. It further stated that the iron used for angles and plates in building this vessel came from Consett, and that the plates were of Brunswick Best, while the iron from LW at Walker was left out for unknown reasons.

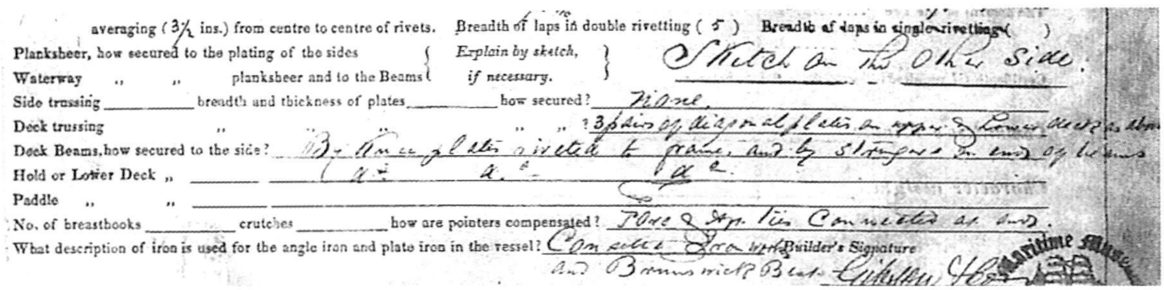
averaging (3½ ins.) from centre to centre of rivets. Breadth of laps in double rivetting (5) Breadth of laps in single rivetting ( )
Planksheer, how secured to the plating of the sides / Waterway " " planksheer and to the Beams — Explain by sketch, if necessary. — Sketch on the Other Side.
Side trussing ______ breadth and thickness of plates ______ how secured? None.
Deck trussing " " " " ? 3 pairs of diagonal plates on upper & Lower decks
Deck Beams, how secured to the side? By Angle plates rivetted to frame, and by stringers on ends of beams
Hold or Lower Deck " ______ do do do
Paddle " " ______
No. of breasthooks ______ crutches ______ how are pointers compensated? Bow & Stn. Tie Connected as [illegible].
What description of iron is used for the angle iron and plate iron in the vessel? Consett Iron Works and Brunswick Best. Builder's Signature [illegible]

The bottom line of the final survey on the Iron Ship *Euterpe* states: Consett Iron Works and Brunswick Best. Courtesy the Maritime Museum of San Diego.

WHAT ARE THESE MARKS?

The next step was to try to identify the names Consett, Walker and Brunswick and establish if they had significance to the building of the ship *Euterpe*. What do the different types of stamp marks actually mean and what information can they provide for us today on the making of iron, the shipment and the ship building of the mid-Nineteenth Century?

CONSETT in County Durham was a wild, desolate place along the river Derwent until 1840, when Derwent Iron Company started wrought iron production at the site. This ironworks strived to produce bars and plates of "first rate quality." By 1860, Consett had become the main producer of plate iron for ship builders at Clydeside as well as at Jarrow, where the famous ship builder Charles Mark Palmer stated that there were no better plates than those from the Consett mills. Even as the leading ironworks in the region, it experienced its ups and downs. In 1863, the original name was changed to Derwent and Consett Iron Company and the year after Derwent was left out of the name (Warren, 1990).

I firmly believe that all of the *Star of India* deck beams for her two decks that carry the stamp CONSETT were produced by Derwent and Consett Iron Co. at Consett.

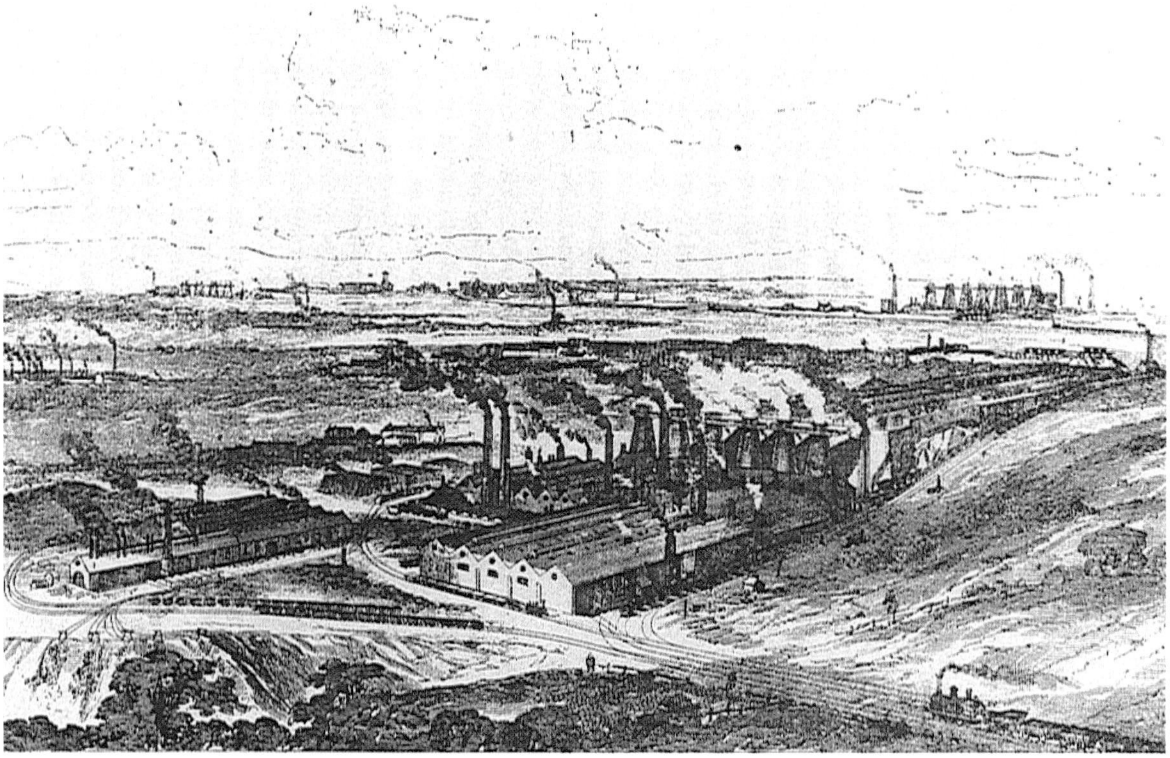

Consett Iron Works at the time iron was delivered to Ramsey for the construction of the ship *Euterpe*. (Warren, 1990, p:54, artist unknown).

Ten years after the Consett mill delivered the iron for the building of the ship *Euterpe* at Ramsey, this company became the biggest producer of ship plates in Great Britain with a production of up to 1,300 tons a week. Consett managed to make the transition into steel production while most old ironworks were shut down after 1880. After 1879, the railway changed to using steel rail instead of iron. Ship builders followed by converting to steel ship construction after ship owners saw that sailing ships of steel could run aground with far less damage than the "softer" ships made of iron. The Consett steel production managed to keep the works in good shape and running well until it was nationalized. When the works finally entered the British Steel Corporation in 1971, it did not take long before tragedy struck. Consett was shut down by Margaret Thatcher's administration in 1980.

LW&WALKER stamps in the frames of the *Star of India* are believed to have originated from the Walker Iron Works on the north bank of the river Tyne, just to the north-east of Consett. This ironworks was founded by William Losh and two young friends — Thomas Wilson and Thomas Bell, next to Losh' alkali factory at Walker. Losh was a colorful and prosperous man who went abroad and engaged in many enterprises, from introducing new methods in soda production to inventing locomotives together with George Stephenson (Welford, 1895). While Consett's iron products were praised for their good quality, the Walker Iron Works did not leave a particularly favorable impression on the supervisor inspecting the different works in the Midlands and the North-East in 1859 (Warren, 1990). The LW & WALKER stamps on the angle iron frames of the *Star of India* represent the names Losh, Wilson and Bell at Walker Iron Works. LW presumably refers to Losh and Wilson. The ironworks was located only three miles east of Newcastle, and provided all the angle iron for the building of the ship *Euterpe*, while all the ship's deck beams stamped CONSETT were produced by Derwent and Consett Iron Co. at Consett. The Walker Iron Works closed down prior to 1910 and the site was developed into a shipyard.

BRUNSWICK BEST mark on the outside hull plating tells us that this plate was delivered from the Brunswick Works. This famous ironworks was located at Wednesbury, a few miles north of Birmingham. The location is 200 miles south of Consett and Walker, but a shorter distance from Liverpool and the Isle of Man than the mentioned ironworks near Newcastle-on-Tyne. Brunswick Iron Works was part of the Patent Shaft and Axel Tree Company at Wednesbury in the Black Country District (T. Round, personal communication, 2003).

In the early 1870s it was described as one of the most prosperous and paying concerns in England. The Patent Shaft Works at Wednesbury was one of the last in the UK to make steel by the Open Hearth steel-making process and was closed down at the same time as Consett. "Best" quality was commonly used to denote iron that was refined twice; telling us that the plate had been processed more than once and thus of superior quality.

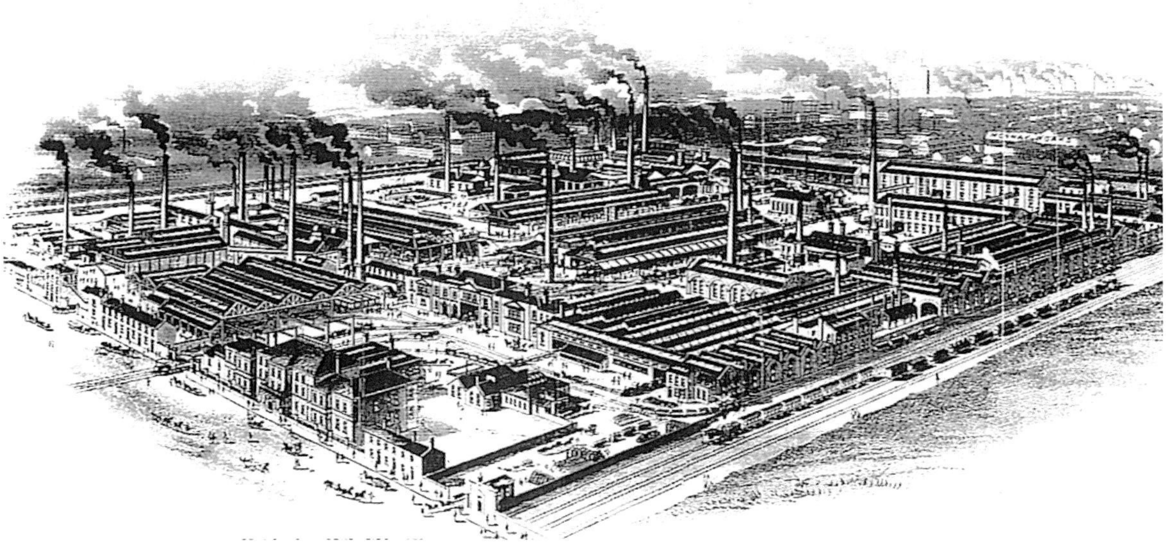

Brunswick Iron Works at Wednesbury, West Midlands, part of Patent Shaft & Axel Tree. The rolling mills are in the foreground to the left, with the furnaces behind. The hammer mills can be seen in the center and the wheel shop at the right. The railroad in foreground and the canal in the back gave sufficient access for shipments, coal and other raw material coming in, iron products going out. Engraving; artist unknown, Courtesy Tom Round.

For the transportation of materials from these three ironworks to the shipyard, the following is the most likely scenario. From Consett and Wednesbury, the iron was shipped by railway to Liverpool and by boat to the yard on the Isle of Man. A major part of Consett's plates and other iron products was sent by rail to the shipyards along the Clyde River in Scotland, about 120 miles to the northeast. Beams or sub-assemblies of 40-foot lengths would not have been any problem for the railways. The Walker Iron Works on the Tyne could have shipped its angle iron directly by sea.

It is important to note that even if I have found stamps from Consett and Walker all over the inside of the *Star of India,* there is only one registered Brunswick imprint amongst all of the hull plating on her outside hull. This imprint, however, corresponds with the final survey of January 1864 that shows that the ship's plates are of the Brunswick Best quality (see page 150).

Oral reports from people participating in dockings of the *Star of India* in San Diego prior to the early 1990s indicate that several impressions with Brunswick Best imprints, or partial remains of this stamp, could be seen on the ships outside hull while cleaning of the ship during dockings. It is therefore the museum's firm belief that the hull plating is still to a large extent, made of Brunswick Best quality iron.

This is supported by my observations during the examination of her inside iron. Riveting repairs performed on the plates and frames usually leave distinct repair-marks and visual modifications to the inside hull. Those are, in many cases, easily spotted as we had learned when restoring the *Hansteen.* I found no such traces of repairs in the *Star of India.* Therefore, I believe that the plates in her hull and bulkheads are her original iron plating delivered to the Isle of Man from the Brunswick Iron Works for the ships construction in 1863.

CONCLUSION

This investigation shows that the old sailing vessel the *Star of India,* ex *Euterpe,* is the oldest existing merchant vessel left in the world, licensed for trade on all oceans, that can still "spread her wings." Most importantly, she is very close, if not, one hundred percent original below her main deck level. Her hull of today is in fact the original wrought iron one. This makes her an amazing and unique relic among the world's existing historic ships. The *Star of India* is a true sailing ship from the mid-Nineteenth Century that has survived very well.

I am pleased to have been able to obtain what I was looking for and help establish the fact that the ship's iron is still where workers placed it when they built the fine ship *Euterpe* on the Isle of Man in the Irish Sea in 1863.

FINAL REMARKS

My studies with regard to boat building, sailing and navigation of historic vessels, and my experience during the restoration of historic ships, outlined in this book, have opened my eyes to the fact that there is little left of what was created in the maritime sector during the first era after the industrial revolution started. Iron was the key raw material in this great leap forward in human history. It helped develop transportation and communication in a scale the world had never seen.

It is a sad fact that few scholars have chosen iron and ship building for research and take it to a level where politicians and public administrators give priority to the protection of surviving samples in general.

Wrought iron has proved to be an extremely durable element compared to other materials. It is nothing less than a disaster when taxpayers' money is spent to replace iron with steel for the pleasure of contemporary sailing, thus ruining any opportunity for later generations to study and admire a real relic of hammer iron that was assembled by hand one and a half century ago.

This book has reflected on new fields this segment of maritime history represents, and it is meant to inspire and motivate others to do something about it. My hope is that this will also spur further research, and help develop a more reflected attitude towards protecting and keeping old iron from being discarded.

Hansteen's engine room. Some 20 of this approved wrought iron hull plates was replaced less then ten years after the ship's complete restoration. Few saw the value of keeping these inferior plates in the ship for modern safety reasons, arguing it is just rusty old iron. Even so, old iron is the key to our world of today.

Near the bank of the River Ouse just south of York, where Viking ships used to sail by for centuries, only two miles from their encampment at Riccall, lies a church constructed not long after the Vikings vanished. The south doorway to this Norman church is a magnificent piece of artwork in stone, wood and wrought iron (Addyman & Goodall, 1979). It includes a commemorative Viking ship with a steering oar. The iron has survived some 850 years, with only minor pieces missing.

Director Peter Addyman of York Archaeological Trust is pointing to the forged Viking ship, part of the decorative iron enforcement of the door to St Helens Church at Stillingfleet. This fine Norman artwork is dated back about 1150 and is an excellent example of how well wrought iron survives if left alone. Since the Medieval Age generations have passed this doorway seeking refuge and support inside the Church.

REFERENCES

Abell, Sir W. (1948). *The shipwright's trade.* Cambridge: Cambridge University Press.

Aberconway, C. (1927). *The basic industries of Great Britain.* London: Benn.

Addyman, P. V., & Goodall, I. H. (1979). The Norman Church and door at Stillingfleet, North Yorkshire, *Archaeologia 106*, 75–105. London: Society of Antiquaries.

Addyman, P. et al. (Eds.). (1981) *The Vikings in England.* London: Anglo-Danish Viking Project.

Aker, R. (1965). *The Cermeno expedition at Drake's Bay 1595.* A Research Report. Palo Alto: Drake Navigators Guild.

Andersen, M. (1895). *Vikingefærden: En illustreret beskrivelse af "Vikings" reise i 1893.* Kristiania: Eget Forlag.

Arlov, T. B. (1996). *Svalbards historie.* Oslo: Aschehoug.

Atkinson, I. (1986). *The Viking ship.* Cambridge: Cambridge University Press.

Bathe, B.W. et al. (1967). *The great age of sail.* Lausanne: Edita.

Brett, Sir H. (1929). *White wings, Fifty years of sail in the New Zealand trade 1850–1900, 1&2.* Auckland: The Brett Printing Company.

Bowen, F. C. (1939). *The flag of the Southern Cross, The history of Shaw. 1858–1939.* London: Savill & Albion Co Ltd.

Brouwer, N. J. (1999). *The international register of historic ships* (3rd ed.). Peekskill: Sea History Press.

Brown, A. C. (1974). *Longboat to Hawaii.* Cambridge: Cornell Maritim Press.

Carlsson, D., & Owen, O. (Eds.) (1996). *Follow the Vikings, highlights of the Viking world/The Viking route.* Visby: Viking Heritage.

Corlett, E. C. B. (1975). *The iron ship.* New York: Arco Publishing.

Crichton, M. (1988). *The 13th warrior.* New York: Ballantine.

Danielsson, B. E. (1964). *Mytteriet på Bounty.* København: Hasselbalchs Forlag.

Dugan, J. (1953). *The great iron ship.* New York: Harper.

Emmerson, G. S. (1983). *S/S Great Eastern.* London: David & Charles.

Engvig, O. T. (1969–75). Arkivalia from interview with fishermen and boatbuilders in Trøndelag 1969–75. Audiotapes, photos, sketches and drawings. Material deposited at Sverresborg Folkemuseum and Norwegian Maritime Museum.

Engvig, O. T. (1977). *Gamle Dampen, Rutebåter på Trondheimsfjorden 1850–1975.* Trondheim: Adresseavisens Forlag.

Engvig, O. T. (1978–1985). Six volumes of logbooks kept on board *Hitra* and *Frøya* during all the 'Viking Way Voyages' along the Norwegian coast, the southern shores of the North Sea to York and the northern route from Norway to England, and across to Great Britain and to the Northern Islands. Not transcribed. Original logbooks in the author's possession.

Engvig, O. T. (1984). I robåt over Atlanteren. *Frivakt, 5,* 10–11.

Engvig, O. T. (1984). Husker du denne?, T/S *Escape*. *Frivakt, 6,* 16.

Engvig, O. T. (1986). Hammer iron and quality marks in preserved 19th century ships, Stamp marks and tests on the hull of sail steamship *Hansteen*. This report is in the author's possession.

Engvig, O. T. (1987). Oseberg replica *Dronningen* and project Vikingfjord. Records on the building and use of the first full-scale copy of the royal barge from Oseberg, Norway, in the author's possession.

Engvig, O. T. (2000). The Viking's heritage. Ships' navigation and migration from the north. *Viking Heritage Magazine, 2,* 35.

Engvig, O.T. (2000). Hammer iron and quality marks in preserved 19th century ships, volume II, Stamp marks found on board the bark *Star of India* built 1863. A preliminary report submitted to the Maritime Museum of San Diego.

Engvig, O. T. (2001). The Viking way. *Viking Heritage Magazine, 1,* 3–7.

Engvig, O. T. (2001). The Viking way. *Viking Heritage Magazine, 2,* 24–28.

Engvig, O. T. (2001). The first shipyard in America. *Viking Heritage Magazine, 4*, 23–25.

Engvig, O. T. (2003). A symbol of kings, The use of golden vanes in Viking ships. *Viking Heritage Magazine, 1*, 3–7.

Engvig, O. T. (2003). A symbol of kings, The use of golden vanes in Viking ships. *Viking Heritage Magazine, 2*, 20–26.

Engvig, O. T. (2003). Evidence in English iron, Stamp marks on *Euterpe*. Mains'l Haul, *Journal of Pacific Maritime History, 39*(3), 4–11.

Espelund, A. (1995). *Iron production in Norway during two millennia.* Trondheim: Arketype.

Espelund, A. (1999). *Bondejern i Norge.* Trondheim: Arketype.

Garyantes, H. F. (1944). *Handbook for shipwrights.* New York: McGraw-Hill.

Gordon, R. B. (1996). *American iron 1607– 1900.* Baltimore: John Hopkins University Press.

Grantham, J. (1842*). Iron as a material for ship-building: A communication to the Polytechnic Society of Liverpool*. London: Simpkin, Marshall, and Co.

Griffiths, S. (1873). *Griffiths' guide to the iron trade of Great Britain.* New York: Augustus Kelley.

Haasum, S. (1974). *Vikingatidens segling och navigasjon.* Stockholm: Esselte.

Hall, R. (1984). *The excavation at York – The Viking dig*. London: The Bodley Head.

Hall, R. (1994). *Book of Viking Age York.* London: B. T. Batsford.

Hanson, B. J. (2001). The Kensington Runestone: Physical features, past and present. *Journal of the West.*

Howarth, D. (1951). *The Shetland bus.* Edinburgh: Thomas Nelson.

Ingstad, H. (1951). *Landet med de kalde kyster.* Oslo: Gyldendal Norsk Forlag.

Ingstad, H. (1969). *Westward to Vinland.* New York: St. Martins Press.

Ingstad, H., & Ingstad, A. S. (2001). *The Viking discovery of America: The excavation of a Norse settlement in L'Anse aux Meadows, Newfoundland.* New York: Checkmaker Books.

Ingstad, A.S. (1994) Letter 9/20 to K. Aartun.

Johnston, I. (2000). *Ships for a nation, 1847–1971, John Brown & Co Clydebank.* Glasgow: West Dunbartonshire Libraries & Museums.

Jones, G. (1984). *A history of the Vikings.* Oxford: Oxford University Press.

Klepp, A. (1983). *Nordlandsbåter og båter fra Trøndelag.* Oslo: Grøndahl.

Larsen, A. (1961). *Åpen båt.* Bergen: J. W. Eides Forlag.

Larsson, M. G. (1997). *Vikingar i Østerled.* Stockholm: Atlantis.

Larsstuvold, T. (1982*). D/S Hansteen, Norges første forskningsfartøy under seil og damp på ny, Vårt Verk, # 5 and #6.* Mo, Norway: Norsk Jernverk.

Lewis, D. (1972*). We, the navigators, the ancient art of landfinding in the Pacific.* Honolulu: University of Hawaii Press.

Lewis, D. (1978). *The voyaging star.* New York: Norton.

Ljone, O. (1953). *Kaptein Toralf Andersen og hans menn.* Bergen: Eides Forlag.

MacMullen, J. (1961). *Star of India, The log of an iron ship.* Berkeley: Howell-North Books.

MacPolin, D. (1999). *The Drontheim, Forgotten sailing boat of the north Irish coast.* Dublin: Play Print.

Mallery, A., & Harrison, M. R. (1979). *The rediscovery of lost America.* New York: E. P. Dutton.

McCarthy, M. (Ed.) (1985*). Iron ships & steam shipwrecks.* Papers from the First Australian Seminar on the Management of Iron Vessels & Steam Shipwrecks, Western Australian Museum, Perth.

Magnusson, M. (1980) *Vikings!* New York: Dutton.

Magnusson, M. (1978). *Liv og Tro i Vikingtiden.* Oslo: Tiden Norsk Forlag.

Morrison, J. H. (1945). *Iron and steel hull steam vessels of the United States 1825–1905.* Salem, MA: Peabody Museum.

Mountfield, D. (1974). *A history of polar exploration.* London: Hamlyn.

Nordhoff, C., & Hall, J. N. (1971). *Mutiny on the Bounty.* Boston: Little, Brown & Co.

Nylen, E. (1982). *I Østerled, Med Vikingaskepp mot Miklagaard, Upstrøms gjenem Polen.* RAGU 1982:2, The Central Board of Nationla Antiquities and the National Historical Museums, Risberg, Udevalla, Sweden.

Nylen, E. (1983). *I Østerled, Med vikingaskepp mot Miklagård.* Visby.

Petersen, P. A. (1966). *Um Mikladals Bygd, Batasmid.* Torshavn: Egid Forlag.

Rowland, K. T. (1971). *The Great Britain.* Newton Abbot Devon: Davis & Charles.

Sawyer, P. H. (1994). *Kings and Vikings, Scandinavia and Europe AD 700–1100.* New York: Barnes & Noble.

Sharpe, M. L. (1931). An apprentice in 1880. *Sea Breezes, 15*(141).

Tames, R. (1988), *Isambard Kingdom Brunel 1806–1859.* Aylesbury: Shire Publication.

Torr, C. (1964). *Ancient ships.* Chicago: Argonaut.

Tweedle, D. (1986). Finds from Parliement Street and other sites in the City Centre. *The Archaeology of York, The Small Finds, AY. 17/4,* 184–186.

Tweddle, D., & Hall, R. (1987). *Viking ships.* York: York Archaeological Trust.

Warren, K. (1990). *Consett iron 1840 to 1980, A study in industrial location.* Oxford: Clarendon Press.

Welford, R. (1895). William Losh, founder of the Walker iron works. *Men of Mark 'twixt Tyne and Tweed', 3,* 92–97.

Wilson, D.M. (1980). *The Vikings and their origins.* London: Thames and Hudson.

Woodman, R. (1997). *The history of the ship.* London: Conway Maritime Press.

Other Resources

Encyclopædia Britannica, Eleventh Edition, Cambridge & New York 1910–1911

Lloyds of London, Lloyds Register, and different Lloyds publications

Newcastle-upon-Tyne City Council, provided information on Walker Iron Works.

Norsk Jernverk A/S (Molab A/S): Three laboratory reports: **MT 69/81** of June 6, 1981 by I. Malin (preliminary test of one item belived to be of Puddling iron). **MT 36/82**, dated March 10, 1982, by R. Nyrud. **MT97/84**, dated July 3, 1984, by S. Johnsen.

The Sagas:
Sæmund Sigfusson, "The Elder Edda" and
Snorre Sturleson: "Younger Edda," Norrøna Society, London 1907
Snorre Sturlason: "Kongesagaer," Stenersen Forlag, Kristiania 1909

The Norse (Icelandic) Ette-sagas:
Hreinsson, V. (gn.ed.): "The Complete Sagas of Icelanders," Reykjavik 1997.

INDEX

Ships' names and Norwegian words are in italic.